# The Binary Firm

## Digital Management Transformation for the Non-Digital Organization

Steven J. Keays, M.A.Sc., P.Eng.

A PRODUCTIVITY PRESS BOOK

First published 2020
by Routledge
52 Vanderbilt Avenue, New York, NY 10017

and by Routledge
2 Park Square, Milton Park, Abingdon, Oxon, OX14 4RN
*Routledge is an imprint of the Taylor & Francis Group, an informa business*

*Library of Congress Cataloging-in-Publication Data*
Names: Keays, Steven, 1963- author.
Title: The binary firm : digital management transformation for the
non-digital organization / Steven J. Keays, M.A.Sc., P.Eng..
Description: New York, NY : Routledge, 2020. |
Includes bibliographical references and index.
Identifiers: LCCN 2020000096 (print) | LCCN 2020000097 (ebook) |
ISBN 9780367900397 (hardback) | ISBN 9780367897574 (paperback) |
ISBN 9781003022046 (ebook)
Subjects: LCSH: Management--Technological innovations. |
Management--Statistical methods. | Big data.
Classification: LCC HD30.2 .K423 2020 (print) | LCC HD30.2 (ebook) | DDC
658/.05--dc23 LC
record available at https://lccn.loc.gov/2020000096
LC ebook record available
at https://lccn.loc.gov/2020000097

ISBN: 978-0-367-90039-7 (hbk)
ISBN: 978-0-367-89757-4 (pbk)
ISBN: 978-1-003-02204-6 (ebk)

Typeset in Garamond
by Deanta Global Publishing Services, Chennai, India

This book is dedicated to our daughter, Gabrielle, and our son, James. She taught me what adversity truly is and how to survive it with integrity. He taught me the meaning of independence of thought and inner strength from conviction. I dare say, because of them, I have become a better human being.

# Contents

## PART III  THE BINARY ORGANIZATION

# List of Figures

# List of Tables

# Acronyms and Abbreviations

**AI**       Artificial Intelligence
**aka**      Also Known As
**AP**       Accountable Party
**AR**       Augmented Reality
**BDaaS**    Big Data as a Service
**BFT**      Byzantine Fault Tolerance
**BIM**      Building Information Management
**BRICS**    Brazil, Russia, India, China, South Africa
**CAD**      Computer Aided Design
**CCD**      Close-Couple Device
**CDO**      Chief Data Officer
**CEO**      Chief Executive Officer
**CERN**     Conseil Européen pour la Recherche Nucléaire
**CIO**      Chief Information Officer
**COO**      Chief Operating Officer
**CMMI**     Capability Maturity Model Integration
**CRM**      Customer Relationship Management
**CSO**      Chief Security Officer
**DAMA**     Data Asset Management Association
**DCAM**     Data Capability Assessment Model
**DCM**      Digital Construct Manager
**DCO**      Digital Construct Officer
**DLT**      Distributed Ledger Technology
**EDM**      Electronics Data Management
**ERP**      Enterprise Resource Planning
**GDPR**     General Data Protection Regulation
**GHG**      Green House Gas
**GPS**      Geo-Positioning System or Global Positioning System

| | |
|---|---|
| **GUI** | Graphical User Interface |
| **HaaS** | Hardware as a Service |
| **HTTP** | HyperText Transfer Protocol |
| **ICIPM** | *Investment-Centric Innovation Project Management* |
| **ICPM** | *Investment-Centric Project Management* |
| **IDC** | International Data Corporation |
| **IEC** | International Electrotechnical Commission |
| **IEEE** | Institute of Electrical and Electronics Engineers |
| **IFL** | Industrial Fabrication Ltd |
| **Igeedee** | Instant-Gratification Device |
| **IIoT** | Industrial Internet of Things |
| **IoT** | Internet of Things |
| **ISO** | International Standards Organization |
| **IT** | Information Technology |
| **ITT** | Invitation To Tender |
| **KPI** | Key Performance Indicator |
| **LHC** | Large Hadron Collider |
| **MBA** | Master Business Administration |
| **MDA** | Multiplatform Data Architecture |
| **MRI** | Magnetic Resonance Imaging |
| **MTBF** | Mean-Time-Between-Failure |
| **MTTF** | Mean-Time-To-Failure |
| **NAIA** | Numeric and Analog Information Assets |
| **NP** | Nondeterministic Polynomial time |
| **OECD** | Organization for Economic Cooperation and Development |
| **OHS** | Occupational Health and Safety |
| **OPEC** | Organization of Petroleum Exporting Countries |
| **OS** | Operating System |
| **PaaS** | Processing as a Services |
| **PAM** | Performance Assessment Metrics |
| **PC** | Personal Computer |
| **PCB** | Printed Circuit Board |
| **PECO** | Project Ecosystem |
| **PLC** | Programmable Logic Controller |
| **PMT** | Project Management Team |
| **PP** | Probate Party |
| **QA/QC** | Quality Assurance/Quality Control |
| **RFP** | Request for Proposal |
| **ROI** | Return on Investment |

| | |
|---|---|
| **RP** | Responsible Party |
| **RTU** | Remote Terminal Unit |
| **SaaS** | Software as a Service |
| **SCADA** | Supervisory Control and Data Acquisition |
| **SWOT** | Strength, Weakness, Opportunity, Threat |
| **TCO** | Total Cost of Ownership |
| **TPM** | Traditional Project Management |
| **UTP** | Unit Transformation Process |
| **VB** | Visual Basic |
| **VDC** | Voltage Direct Current |
| **VR** | Virtual Reality |
| **W5H** | What, Why, When, Where, Who, and How |

# Preface

*A fad isn't a fad when it's no longer a fad.*

"All art is quite useless". Thus spake Oscar Wilde in his preface to *The Picture of Dorian Gray*. Wilde went on to explain his famous declaration by juxtaposing the uselessness of art against the uselessness of a flower. A flower simply is, without reason or purpose. It is in the act of observing it, of appreciating it, of caring for it that a meaning, if not a purpose, percolates to the surface of our senses. Wilde was indisputably right in regarding art's aim as "simply to create a mood". Waxing poetic in the context of a book whose nature is, by its very name, technocratic, may strike the reader as unexpected, to say the least. Nevertheless, it was this quote from Wilde that inspired this book, the third installment of the investment-centric series initiated in 2017 with the publication of *Investment-Centric Project Management*. That, coupled with the intriguing signs that were sprouting, also in 2017, in the popular, academic, and business literature about the advent of powerful software everywhere. Software that mimicked human intelligence and inference, that was capable of driving cars, of animating nightmarish humanoid robots (from the geniuses at Boston Dynamics), and even of catching people in the act of lying. The prospects of this new digital machinery began to sweep across corporate boardrooms and governments alike. Venture capital was quick to join the party; it continues to hunt far and wide for the next killer app. Simultaneously, books began to appear in rapid order and painted various landscapes of this new art of the digital. Their form follows the classic script of past novating movements such as the art of management, of quality, of engineering, of the deal. Some of these books have yielded important insights that have found their way into this one, especially *Data Strategy* (2017, Kogan Page) by Bernard Marr and *The Fourth Industrial Revolution* (2017, Crown Business) by Klaus Schwab.

It would seem, generally speaking, that books on the art of the digital, for lack of a better expression, excel at describing the new digital world and the promises it holds for those who hop on the train. One can only describe, comment, and admire the Art of anything, since the artist is the only one possessed of the skills to create. The same goes for the current crop of books on the digital era (which will be referred in this book as the *Symbiocene age*); readers are encouraged to hurry up and join the party for dibs to the choicest pieces off the buffet table. This weak form of proselytizing is reminiscent of the 1980s management fad to "work smarter, not harder", focusing always on describing all that can be smarter but never *prescribing* how to do so. The digital transformation message resonates intrinsically well with large, digitally savvy organizations with a built-in capacity to undertake the journey. Unfortunately, this niche audience excludes the vast majority of other businesses and organizations that go about their daily affairs with varying degrees of digitalization. Hence, this overwhelming majority is effectively shut out of the conversation, be it on account of insufficient scale, primitive numeric literacy, or superficial relationships with the internet. For these organizations, being told to "work smarter, not harder" as they contemplate the coming digital revolution amounts to naught.

The impetus for this book was, and remains twofold: 1) *prescribe*, in concise yet actionable details, the means of undertaking one's own, bespoke digital transformation; and 2) target the silent majority of organizations that stand to benefit from such a transformation, regardless of their starting point. Getting to the "working smarter" part is done quickly at the book's outset. The rest of the text will dwell on the mechanics and mechanisms of doing so. The reader will find practical, sensical techniques, methods, and tools for figuring out an organization's current digital acumen (or lack thereof); for determining what makes sense for this organization (as opposed to buying into the latest whizbang technology just because everyone's talking about it); and how to plan, strategize, and implement the digital solutions that will permanently transform the organization, *always with the overriding objective of maximizing the organization's long term return-on-investment performance*. As a matter of fact, the reader will find a recurring caution throughout the text against blindly embracing the latest and greatest. Instead, the text is orchestrated along a linear progression from current state to opportunities then to business cases and finally, to finding and implementing the most suitable solutions for satisfying the business cases.

Finally, the human element in all this human maelstrom is kept front and center to all aspects of a digital transformation journey. Again and again, the reader will hear that the key to succeeding in this journey isn't about picking the right technology; it is about molding the human-machine interface into the best synergy possible to make the former excel by making the latter perform. It is why this book speaks of digital *management* transformation, rather than the more common expression *digital transformation*. In the Symbiocene age, neither can thrive without the other.

# Acknowledgements

*Third time's a charm.*

This third installment of the *investment-centric* series came about through osmosis with my professional interactions over the past two years. As before, my wife Margaret was instrumental in creating and maintaining an environment in which the selfish act of writing can flourish, which speaks volumes about the special person that she is. It behooves me to reiterate the sentiment expressed in the previous two books: *I would not be me, without you with me.* She also lent her artistic touch to the creation of Figure 3.1 jointly with Agnieszka Bak.

This book would not have come to public life without the meticulous work of Mike Sinocchi and Katherine Kadian at Taylor & Francis. The dynamic duo is a writer's best friend!

The final round of encomia goes to several business associates who contributed precious insights from their sphere of influence, such as Brendan Boyle (Vintritech.com), Jody Conrad (Kruxanalytics.com), Dave Houston (Miracad.com), Pawel Pisarski (Lumiant.com), Peter Guo (www.whitewhale.ai), and Charles Robison (Coredata.ca).

Lastly, I would be remiss to not tip my hat to Fraser Brooks and his crew at Starbucks, Britannia Plaza (Calgary, Alberta), who kept the brain lubricants flowing, by the bay window... I raise my coffee cup (my very own, really, kept beneath the coffee machine cupboard) to Ainna Javier, Cale Black, Cheyenne Ockey, Gemma Maglaque, Gregory Vanderbeek, Heather Reid Rickard, Ismam Rahman, Justina Rogalsky, Katherine Kaye Moon, Kim Nursall, Lucille Fisher, Montana Albrecht, Pearpreet Kaur, and Yenimar Badell. Business must be good: the list is longer than before...

# Author

**Steven James Keays, M.A.Sc., P.Eng.,** is an international-selling author, keynote speaker, management consultant in major projects, modularization and construction, and technology development. He is a 35-years veteran of the aerospace, defense, energy, manufacturing and Oil and Gas sectors. He is a professional engineer, a graduate of the Royal Military College of Canada, founder and CEO of both *NAIAD Company Ltd* and *The Institute of Advanced Management*. Mr. Keays has written extensively on project management, innovation and technology, and digital strategies. His published works include *Le Big Bang Canadien, Investment-Centric Project Management* (2017, J. Ross Publishing), *and Investment-Centric Innovation Project Management* (2018, J. Ross Publishing). Mr. Keays and his wife Margaret are proud parents of three children, Michael, Gabrielle, and James, and resides in Calgary, Alberta, Canada.

# Introduction:
# Chapter Summaries

The structure of this book is thematically divided into three parts. Part I comprises the first three chapters. The reader is introduced to the nascent digital universe that is rapidly transforming our world, especially where business is concerned, and the consequences that are starting to be felt far and wide by individuals, people, organizations, and governments alike. Part II outlines the four stages of a digital management transformation, from scoping to implementation. Each stage is explored in a chapter, which proceeds to present pertinent objectives, tools, techniques, and methods in play. The four chapters follow the same arc of execution that is followed when the transformation journey is undertaken by an organization. Part III is dedicated to the means and measures of managing the transformation journey, with an emphasis on the human variable. The structure of this book is chosen to maximize the reader's rapid grasp of the topics presented, and to achieve a rapid pace of execution through an *adaptive strategy* tailorable to the unique circumstances of the reader. The guiding imperative of this book is to equip the reader with the wherewithal to achieve enhanced returns on investment rapidly, without breaking the bank.

## Part I: Terra Digita

- Chapter 1, *Introduction*, sets the stage for anyone contemplating a digital transformation, usually of an organization that is far from having achieved technological mastery of the hardware and software heralded everywhere as building blocks of the new industrial revolution.
- Chapter 2, *The Symbiocene Age*, segues to the revolutionary theme. It makes the case that we are indeed witnessing a new industrial and economic age unfolding before our eyes. Parallels with the upheavals

of past centuries are highlighted, starting with the first industrial revolution. The text introduces the reader to the concept of the *holobiome*, a neologism coined in this book to capture the sinews of the new digital landscape and its ramifications upon the way business is conducted.

■ Chapter 3, *The Nature of Data*, takes the reader into the murky world of data, a word that has become so ubiquitous as to have lost its meaning (i.e., the plural of datum). The text pivots toward the idea of data as financial assets able to generate revenues; it introduces the concept of the *naia*, the neologism coined to represent the data, information, knowledge, and intelligence holdings of an organization; presents a comprehensive but practical methodology for quantifying and qualifying the individual instances of the organization's *naia*; and takes the reader through the challenges and perils of handling data without a viable digital management framework.

## Part II: The Journey

■ Chapter 4, *Data Demands*, sets the starting line of the digital management transformation journey. The text focuses on the organization's *naia* as it exists now, in what format, and to what purpose. The goal is to establish an information *baseline* upon which the digital transformation is to operate. The concept of the *analog opportunity cost* is introduced as the hidden costs of the organization's status quo through its embedded limitations, inabilities, and ROI-destroying features. The other foundational concept introduced in Chapter 4 is the *digital construct*, which is the operational system under which the hardware, software, user interfaces, and procedures are managed. At the conclusion of the chapter, the reader can expect to have an understanding of the totality of the organization's information landscape.

■ Chapter 5, *The Business Model*, takes the reader through the process of mapping out a viable transformation path from the information baseline to a digitalized organizational framework able to make money from its *naia*. The process leads one to define the *model calls*, which stipulate what commercial value is to be ascribed to every instance of the *naia*, and specify what outcomes are to be achieved when the digital transformation has been applied to these instances. The *model calls* are, in effect, the strategic targets of the organization. Finally, the reader is introduced to a pair of neologisms, the *plik set* and the *numer set*, to differentiate the *naia* instances that are handled primarily by humans, and those handled by machines, devices, and algorithms, respectively.

■ Chapter 6, *Binary Tools*, instructs the reader on the methodology for choosing the digital technologies that can best give form and substance to the *model calls*. At this stage of the transformation journey, the reader is moving from defining the organization's needs and requirements to specifying the technological requirements and specifications of what will eventually be bought and implemented in the operational setting. These specifications form the basis of selection for the various components of the *digital construct.*

■ Chapter 7, *Engagement*, takes the requirements and specifications of Chapter 6 and converts them into buying decisions. In Chapter 7, the organization begins to roll out the physical aspects of the *digital construct*, initially with a pilot program then throughout the entire organization. By the end of the chapter, the reader's organization will be operating, in real time, in a newly forged digital setting.

# Part III: The Binary Organization

■ Chapter 8, *People Digital*, changes perspective, with technological considerations giving way to issues of management. An argument is put forward to the effect that the success of a digital transformation is *impossible* without integrating the human variable into the very fabric of the new digital setting. The text explores the salient impacts of the new technologies upon the nature of the work by employees, contractors, and supply chain partners alike; discusses the imperative to train one's workforce before things are rolled out; and highlights some of the new job types that come with this new technological framework.

■ Chapter 9, *Project Digital*, outlines a case study whereby an organization involved with large industrial capital projects can look to its project delivery framework as the ideal candidate to carry out a digital management transformation *without affecting the rest of the operations.*

■ Chapter 10, *Conclusion*, rounds things out by highlighting the fact that the Symbiocene age is already happening all around us; survival of the fittest passes necessarily through a digital management transformation. One that can only succeed when the implementation strategy is anchored solidly to the human-machine symbiosis to extract all that can be positively extracted from this new frontier.

# TERRA DIGITA 1

# Chapter 1

## Introduction: A Brave New World

*Don't fight the problem; decide it.*

**General George Marshall**

Every firm and organization operating in the world today are either connected to the internet or handling digital transactions in one form or another, or both. A fraction of them requires a digital strategy in the traditional sense (in which mobile connectivity with customers drives business). For the overwhelming majority that remains, what they actually need is a digital management strategy to transform their relationships with data in a way that will increase their profits. This book is for the latter.

## A Preliminary Perspective

### Columbus' Fifth Voyage

The world of business, as we know it today, is on the cusp of a permanent transformation into a reality that has yet to solidify. Much like the world that Christopher Columbus* left, on 3rd August 1492, would never be the same upon his return on 4th March 1493, the world that we inhabit today will be vastly different in 2022 and beyond, and simultaneously unforgiving and rewarding. The symbiotic convergence of data, machines, software, people, and organizations is rapidly shaping a new commercial landscape without equal in history. The transformation is chaotic, viral, relentless, and irreversible. It is taking place in real time, under our very eyes, at speeds that defy control. No aspect of society is being spared: witness the legal travails of Uber and Facebook, to name but two, against the forces of entrenched competition, government interventions, and privacy concerns. Even venerable legal systems, hitherto inured against the pressures of modernization and innovation, are having to wheel and deal their ways into ugly, uncomfortable quagmires spurred forth by legitimacy issues surrounding cryptocurrencies, data ownership, overbearing governments, and public information flows. Klaus Schwab, founder of the World Economic Forum, called this transformation the *fourth industrial revolution.*[†] This effervescent period in human history is more than a revolution however, given the arrival of machines unto the human conscience. The period is deserving of a distinct title: the *Symbiocene Age*, whose construct will be explored in this book as foundational.[‡] This neologism is formed by the contraction of the words *symbiosis*, representing the conjoined existence of humans and data, and the suffix *cene*, denoting a geological period. The age is bearing witness to the unstoppable ascendency of digital data as drivers of the transformation. The very expression "digital

---

* Columbus made four voyages to the new world: 1492, 1493, 1498, and 1502 (see Bergreen: *Columbus – The Four Voyages*, p. 4 [2011, Viking Press]). His journey traces its roots back to Palos de la Frontera in southern Spain—discovered in 2014 as the genuine launching port of Columbus' first expedition. It would seem strange that the actual location of the departure would be still subject of archeological research. However, the findings of archeologist Juan Manuel Campos, of the University of Huelva, lend credence to this conclusion. Historical records indicate that the La Fontanilla port had a shipyard, pottery works, a freshwater fountain, and a reef that served as the port's customs area. It would also have been the place where Columbus negotiated the logistics of the enterprise. Source: *Archaeology*. 2014. Location of Columbus' point of departure found in Spain. Available at: www.archaeology.org/news/2575-141007-spain-columbus-port.

[†] Schwab's excellent book *The Fourth Industrial Revolution* (2017, Crown Publishing Group) offers a superb overview of the extents and ramifications of the digital wave upon the world.

[‡] The neologism *Symbiocene* was independently coined, somewhat obscurely, by Glenn Aalbrecht in 2011 but with a dramatically different, ethereal meaning. Aalbrecht conceptualized his *Symbiocene* as an ecosystem characterized by interconnectedness of all living things and governed the categorical imperative to make all decisions pursuant to the greater good of Nature. Aalbrecht's formulation excludes technology and digital systems altogether, whereas they are foundational to the definition embraced in this book.

transformation" has already achieved buzzword status in the public's mind. Satya Nadella, CEO of Microsoft, gave impetus to this transformation by declaring in 2015 that every business will become a software business in the future.* Any business today, which transacts with customers principally through the internet or a mobile device, has no choice but to go all in with its own digital transformation, or perish. Going digital is generally viewed across corporate boards as a binary proposition: either join the digital bandwagon or be left behind at your peril. There is no halfway, no fence straddling, no tepid intent allowed. It is all in or all out. Survive or die or wither away in painful agony.

This black or white choice inspired the title of this book. While the injunction to become a software company is not as absolute as it is purported to be, there remains in it a kernel of truth in that the Symbiocene age will rapidly weed out laggards and *status quo* huggers alike, and richly reward those who aggressively heed the call to digital arms. Businesses must either get on board with the new reality or face the threat of accelerated senescence. The consequences of strategic inertia ("let's wait until the dust settle to see where we should go"), of illusions of uniqueness ("our business doesn't need to transform because our business model is so unique"), and of defending the status quo ("calm down boys, this is not our first rodeo riding the 'it' fad of the moment") will be equally binary in outcomes. You will vanish quickly, rather than drag things along for years. You will not be given second chances: once you lose out to the digitalized competition, you are done. There is no middle ground of commercial survivability. Remember Kodak? Remember Sears? The blatant facts of the new age are simple: to thrive in the Symbiocene age, a business must become binary across its own landscape. Binary at the information level; binary in joining people and data across business processes; binary toward its management philosophy (people on one side, software on the other); and binary in embracing the digitalization of one's inheritance. There is no room for gray areas, for hesitant adoption, for shades of strategic options. The Symbiocene age is upon us.

---

* Nadella uttered the famous phrase when addressing delegates to Microsoft's 2015 US conference (March 16, Atlanta, Georgia). Nadella said: "Your drive to transform businesses is setting the technology agenda. What you want to achieve in business is fundamentally changing technology". Nadella also observed that the coming internet of things (not yet a "thing" back in 2015) promised to affect all businesses by making them software businesses. Nadella urged the audience to look at their respective businesses through data lenses to guide their transformations because "every business will become a software business, build applications, use advanced analytics and provide SaaS services".

Resistance is futile, as the Borg would say.* Time to get on with the business of digitalizing your business.

## LANGUAGE MATTERS

It is important to distinguish between datum, data, digitize, and digitalize. The word "data" is grammatically the plural form of the singular "datum". Decades of vernacular preferences have made "data" into a normative form as well, such that it is understood to be singular according to circumstances. For instance, one should, grammatically, really refer to a "datum base" rather than a "database". Nevertheless, the ubiquitous usage of "data" renders all argument against its usage futile. Hence, the convention for "data" will be (reluctantly!) adopted in this text. The word "datum" will be utilized in the specific instance of expressing a singular piece of information from a bigger set.

Two conventions will be used to distinguish the verbs "digitize" and the neologism "digitalize". The former will be used according to its traditional definition in a technological context: that is, to transform an input into a binary sequence of zeros and ones (scanning an image from paper to a jpeg format, for example). "Digitalize" will be used to describe the conversion of an analog process into a *digital construct*. For example, a museum seeking to digitize ancient manuscripts will go through a digitalization effort that includes the purchase of scanners and database and management software to scan, capture, convert, store, track, and distribute the datasets obtained from scanning the pages of old books. The "digital construct" comprises the equipment, the computer hardware, the datum streams, the datasets, the software/algorithms, and the user interfaces to execute the digitizing process and manipulate/transact the resulting digitized data.

If a thing or process cannot interact with a software algorithm, it is analog (books, door keys, signatures, welding, trucking, cooking, etc.). Essentially, our modern world is still an analog world when you look at everything that is still done without any computer help.

The most important element of the "digital construct" is the algorithm, which is the computerized set of instructions to transform a set of digital inputs (the data) into another set of digital outputs. Digital data are both

---

* From the *Star Trek: The Next Generation* series. The famous declaration by the Borg, uttered to would-be conquered nations, when like this: resistance is futile. You will be assimilated into the Borg collective.

utterly useless and wasteful if they cannot be transformed into useful outputs.

Finally, the word "wart" will be used figuratively to refer to something (an object, a process, a procedure, a decision, etc.) that is done a certain way merely because it has always been done that way. The wart is the manifestation of an entrenched *status quo* which has never been questioned before. The wart is also the greatest opportunity to improve through its eradication. *The instant when the status quo needs to be justified is the time when it is no longer justifiable.*

The transformation under way gave spawned the Symbiocene age; but the age is more than the transformation. While the effects of the transformation are immediately visible to the general public (incarnated by the ubiquitous smartphone), its ramifications operate at the *mercantile* depth, where businesses, corporations, regulators, and governments exist. The general public is constantly at the *receiving end* of every newfangled gadget or service introduced in the marketplace; the *generating end*, on the other hand, is firmly held by the business players who create them in the first place. Granted, the general public is locked into an adversarial feedback loop with the creators. if only to evolve* the features on offer into winners and losers. It is indeed the process that killed off Apple's "Newton" personal digital assistant but crowned the company's "iPhone". But this feedback loop is of secondary importance to the complex pursuit of innovation under the aegis of profits. The inchoate, some might say unbridled, explosion of all things digital upon the retail, public, and business worlds is spurred on relentlessly forward by an insatiable thirst for convenience (at the *receiving end*) coupled to a bottomless quest for riches at the *generating end*. There is no way to predict how the world will look even two years out, such being the speed of the transformation. Nor is there a way to pinpoint a logical "end point" to the transformation—echoing once again the same evolutionary mechanics of natural selection. Columbus could never have had the intellect to fathom what would come out of his fourth and last voyage to the new world. Our Symbiocene age is akin to a fictitious fifth voyage by the famous Genoan explorer. There is no way, no conceivable way to predict what the world will be two, five, or ten years hence. There are, nevertheless, three predictions that can indeed be made without the augury power of soothsaying: 1) the

---

* The word "evolve" follows exactly the meaning defined by the process of "natural selection".

transformation will continue unabated, impervious to all antagonistic oppositions; 2) those who refuse to embrace this new world will be left to wither away in the old; and 3) those willing to embark on its journey will the first to lay claim to the riches that are yet to be discovered in this new country.

## A Modern Mythical Spice Trade Route

Columbus stumbled upon a new continent out of sheer luck and misplaced greed for finding a new trade route to the spice riches of Asia. We are much better placed, in our Symbiocene age, to find our own roads to riches. The belligerent competition among nations in Columbus' times is equally active today, save for the attendant wars. The marketplaces within which the battles for digital supremacy are being fought are no less foreboding than their historical ancestors. For, make no mistake about it, the digital transformation is a genuine war of competing beliefs and self-interests, one which will take few if no prisoners. For any business aspiring to grow and thrive, regardless of its particular market, the choice is, once again, binary: either join the fight and win, or avoid the fight and suffer the ignominy of conquest.

### EVEN WHEN HE WAS WRONG, HE WAS RIGHT

There is a delicious irony with the fact that the pitch that Columbus made to kings and queens to finance his first expedition was based on a gross under estimation of the travel distances involved. Recall that Columbus was motivated by the belief that the East Indies (today's South and South East Asia) could be reached faster by sailing westward, thereby establishing a lucrative spice trade route superior to all others heading eastward. He estimated the westward route to be about 3700 km (in modern parlance), based on the equivalence of 56.7 miles per degree of latitude, derived earlier by an Arabic scholar, Ahmad ibn Muhammad ibn Kathir al-Farghani (aka Alfraganus, 9th century). Afraganus had used *Arabic* miles as units; Columbus interpreted them in *roman* miles, which were shorter. This error in units was compounded by other estimates of distance made by various European at the time who located Japan much closer to Europe than it actually was. The correct distance was six times greater.*

---

* See *Admiral of the Ocean Sea: The Life of Christopher Columbus* by Samuel Eliot Morison (1942, Atlantic/Little, Brown; Reissued: 2007, Morison Press).

Had Columbus used this, his investor pitch would have fallen on deaf ears: no ship of that era was believed capable of traversing such a distance without running out of food and water. The maritime logic of the time argued arrantly against the project. Hence, it was greed and war debts that won the day from the regents of Catholic Spain (bankrupted by ceaseless wars on the continent). The irresistible allure of profits from such a novel trade route cast aside all counter arguments. Thus was Columbus allowed to cast out on his fateful journey. Never underestimate Mammon!

Fortunately, not all is doom and gloom for organizational leaders. Indeed, the upside of the coming age far exceeds the downside, at least for the businessperson intent on turning the turmoil to his or her benefit. For starters, the prospect of digitalizing one's business must be regarded as a rare opportunity to make a deep dive into its inner workings, to figure out why things are done the way they are done and why in no other way; to uncover the inefficiencies, the bottlenecks, and the warts; to assess what is missed, invisible, or unknown; and to discover new ways to "monetize" what has hitherto been invisible to the bottom line. The undertaking also offers the business a unique opportunity to engage its customers and supply chain partners beyond the mere transactional, to find out how those relationships can be improved, enriched, shared, and strengthened through the power of digitalization. In this way, the sedulous manager will have obtained the first quantified picture of the business' *analog opportunity cost*\* (addressed in Chapter 4).

There is more good news, stemming from a cleansing of widespread myths associated with digital transformations. These myths are promulgated as inevitable truths by a self-serving coterie of consultants and management "gurus" vested in their own digital future. Some myths will apply, to a certain extent, to businesses whose profits are heavily dependent on a constant tethering of their customers to their internal revenue-generating activities (social media, stores, retail operators, etc.). But if the business does not interact with individual consumers, these myths turn out to be misleading. In effect, if your business can be qualified as an industrial operation, in the broadest possible sense of the term, these myths are NOT for you. For instance:

---

\* The opportunity cost is a traditional financial expression defined as the loss of future potential gains from choosing a specific path among two or more options. One can invest $1000 in a government bond, a corporate bond, a share purchase, or a private loan. Either one defines the known upside for the investor and the downside of forgoing the upsides of the other three options. The same goes for the downside, incidentally. Picking the government bond guarantees a specific return but writes off any possibility of higher earnings of the other options.

**You must become a software company**. This is a myth, pure and simple. You have absolutely no need to become a software company. The core expertise of your company will remain your core once you are done with your bespoke transformation. What is required of a business is to figure out what specific elements of the digital universe make sense to your core and how to integrate those elements into an expanded *business model* designed to monetize the new paths to profits and profitability.

**Your business model must be anchored to a data strategy**. You are not needing to become a data company either. You will need some kind of a strategy about data; that is inevitable. Which means how to create value out of one's business datasets. But there is no need to make the case for transforming every function of the business into a generator or consumer of digital data. Data is subservient to the business, not the reverse.

**Disruptive technologies will anchor the transformation**. As a matter of fact, disruptive innovations can be ignored. First, the expression is grossly overused to describe what are functionally incremental in most cases. The internet was disruptive. So were the CCD camera, the iPod, the iPhone, and the cloud as software delivery platforms. But self-driving cars, Uber, Airbnb, and whatever else in the business literature peddled as disruptive, were not, in fact. Second, the low hanging fruits in any digital transformation are found in the internal mechanics and mechanisms of a business. Innovation by improvements and refinements is more potent and more immediately propitious to the bottom line than whizzbang gadgets and shiny new phone apps.

**The digital transformation must be companywide**. Another myth that is harmful in many instances. A digital transformation is best done specifically rather than ubiquitously. Most businesses that are not involved with constant digital interactions with consumers have no need for changing every facet of their operations into a digital framework. The most bang for the digital buck will be obtained when the business determines what analog processes are ripe for digitalization or the need for digitalization to eradicate the warts of the underlying status quo.

**The transformation must be baked in from the get-go**. This is the corollary to the preceding myth and is equally deleterious. A successful digital transformation proceeds from small steps, limited in scope and corralled in risk. And, do not forget, going in gradually minimizes the demand on cash flow.

You do not need to embrace an overarching mobile strategy. You do only if your profits are made or destroyed from direct, mobile interactions between your customers and your operations.

**You need to transform now**. Contrary to popular belief, and despite the speed of the global transformation, you have in fact the luxury of manageable time to develop a strategy to develop the digitalization of your business. Where there is urgency is in the recognition that this transformation is necessary. The Symbiocene age is here; you have no choice in the matter. You must therefore get on board this train now. But you have the ability to dictate which tracks to ride, the time of departure, the speed of travel, the stops along the way, the target destination, and the options to alter course if you need to.

**You must disrupt your industry before it disrupts you**. Again, the worn usage of the term disruption belies the fallacy of the urgency. Very few businesses have the heft, the wherewithal, and the financial resources to lead the disruption charge. And fewer still should attempt these "moonshot" peregrinations. To put it simply, chasing the bleeding edge, come hell or high water, always burns through mountains of cash, capital, and patience, and without even any assurance that the direction sought will become the path well-travelled. The better, safer, and capital-preserving approach is to keep an eye on what is genuinely trending vs what is just a fad, then extract the best of both and deploy it within your circumstance.

**Software solves all**. One cannot speak of digitalization without of course speaking about software. It is self-evident that software is integral to the process. But it does not follow that software is in the lead. No amount of supreme *artificial intelligence* application can magically save a business whose *modus operandi* is opaque, inefficient, archaic, or stubborn. Software never is the answer. Software is the means of manifesting one's correct understanding of how the business functions (or not). Software is a tool, not salvation.

**A digital transformation belongs to the IT department**. Patently false. The transformation is first and foremost a corporate board endeavor, delegated to the CEO or the COO to implement. The IT group will be involved in the plumbing and the hardware.

**A digital transformation is about cost and labor cutting**. This objective is the worst possible reason why any business should undertake a digital transformation. Process, procedure, and job casualties are quite likely to happen, for the simple reason that the *analog opportunity cost* will include inefficiencies that can be remedied with better processes, automated machines, and reduced head count. But it can also include untapped monetization opportunities that will require new equipment, new procedures, and new people to implement. Additionally, the human variable is the invisible vector of success or failure to any transformative program, as we will see in Chapter 7. Robots, software, and complex IT systems are, on their own, the fastest way to financial meltdown; a consequence of the first digital paradox, described next.

**Transformation is about technology**. Naturally, any digital transformation implies, *inter alia*, technology (in the widest possible sense). Technology, as a product or a schema (see capsule below), is foundational. There can be no digital transformation without it. Furthermore, the economic potency of any transformation is directly linked to the embedded analytical capacity of the many elements making up the technology infrastructure. This capacity concept is explored in Chapter 4. But it is a myth to believe that technology alone will achieve lasting results. To be truly transformative and profitable, technology requires human interactions. That is the first digital paradox of any transformation endeavor:

Paradox 1: Technology needs humans to succeed.

## MORE DEFINITIONS

The terms "product" and "schema" were defined in *Investment-Centric Innovation Project Management.** "Product" encompasses both product and service. The product is bought to perform one or more functions required by the buyer. It can be physical or algorithmic, and exhibit either mass or energy consumption or both. A service, on the other hand, has neither mass nor consumption. A service is an activity that produces an output that can be bought or used. From the user's perspective, a product answers the question "what can I buy from you" whereas a service answers the question "what can you do for me". Clearly, the inner workings of a service may involve physical and algorithmic products.

Products are divided into three types: *integrated*, *ready*, and *schema*. The first two are again physical or algorithmic (a circuit breaker and a shoe, respectively). The *schema* product has no tangible form. It is first and foremost a process through which information and interactions flow. The *schema* product is akin to the mechanics of a transaction. What must be done to approve an expense report, for example. It exists as a cog in the wheel that makes an organization's world turn. Notably, the *schema* product offers the largest collection of low hanging fruit for immediate benefit from digitalization.

**Automation leads to perpetual efficiencies**. The corollary to the cost cutting myth concerns productivity and a second digital paradox:

---

* The meaning of the terms "product" and "schema" follows the convention adopted in Chapter 1 of the book *Investment-Centric Innovation Project Management*.

Paradox 2: Profitability decreases, over the long run, with increases in automation.

The first time out, a widespread automation program produces material gains in throughput efficiencies, unit cost reductions, and aggregate profit increases. But automation by one competitor inevitably leads to an economic arms race by the rest of the field. Once the entire field is nominally equalized, the room for further gains become ever slimmer. The sunk capital costs of an automation program make it exceedingly difficult to justify tearing things up to get new gains. Automation, in other words, leads to competitive stagnation while simultaneously killing the prospects of new innovations that are unique to human eyes (and expertise).

# A Brief History of Then

## *The Existential Business Questions*

If we admit the plausibility of a fourth industrial revolution—and any remaining doubts will be cast away in the next chapter—we are left with four pressing questions:

- How does a business goes about transforming itself amidst the onslaught of all things digital? (aka the *internet of things* [IoT], artificial intelligence [AI], blockchains, autonomous systems, cryptocurrencies, material design, instant manufacturing, and all the rest that glitters);
- How does one *manage* the people and machines without killing the essence of the place?
- How can a business cope with the unrelenting threats of future technological unknowns without needing to constantly change the business model?
- How to make the business not only withstand the future but profit from it now?

The future is already here, inchoate yet already disruptive. From a business strategy perspective, nothing about this is obvious. Finding the answers to the above questions can be overwhelming; but it does not have to be. These answers are woven into the fabric of the remaining chapters.

The reader may have picked up on a pair of missing questions: How to select a specific digital solution and how to implement it? These are technical questions delving into the minutia of the weeds. The literature is already awash with competent texts too numerous to explore within the scope of this book. Starter titles are listed in the bibliography. The important point to make at this juncture is this: one must first master the management of a digital transformation (what this book is about) before introducing new software and hardware solutions into the *modus operandi* of the business (those other books). Recall the implications of the first digital paradox:

> Paradox 1: The key to success in any digital transformation lies with the human variable, with technology playing a supporting role. And the key to success in any human equation is competent management, not processes and procedures.

## A New Hegemon

However, there is one more primordial question that must be addressed before all others: is it really so important to hop on the digital bandwagon? After all, many readers may be employees of a business that has been around longer than the smartphone or the internet. Longevity implies a successful history of adaption by the business to the many fads that came and went over the years. Having survived thus far, why would this time be any different? Why not just go with the flow, as in the past, and integrate *mature technologies* when the time is ripe? If you lived to tell the photocopier marvel of the 1960s, the fax machine paradigm of the 1970s, the quality discourse of the 1980s, the office automation of the 1990s, the internet connectivity of the 2000s, and the mobile magic of the 2010s, why would this business of digital hegemony be treated any differently? Why not just buy the hardware as a matter of business routine and install it? The answer, in short, is a resounding, violent "yes, it is that important". The Symbiocene age is not simply a heteroclite grouping of fancy, newfangled technologies or novel business concepts. It is, quite literally, a radical departure from past history. Whereas until now, processes and technologies and management principles were discrete, independent building blocks put together in ways that created business models as unique as the businesses themselves, the digital paradigm that is imperviously making its way into the human world cannot be adopted or managed through the sum of its parts; the sum itself dictates the nature of the parts—humans, machines, and schemas—as well as their mutual interactions. The causation operates in reverse, from sum to parts. That is why it is important to get on the bandwagon. Buying bits and pieces and machines won't cut it. There is much more at stake.

The consequence of this causation reversal is already visible. The business world has become more competitive, more demanding, more integrated. Beyond 2025, it will be brutal. Technology will be its undisputed master. It is a funny thing, this technology chimera. It is both enabler and destroyer, both wealth generator and destitution giver. Its skeleton is digital, its mind algorithmic, and its blood data streams. Much like its make-up, it is binary in every aspect: on or off, succeed or fail, thrive or perish, wealth or poverty, in or out. Technology is a ruthless invader uninterested in taking prisoners. But technology can be a benevolent tyrant, blind to whomever it favors, welcoming of new petitioners. It is wildly generous to the brave, the curious, the agile, but intolerant of the complacent, the staid, the cautious, or the defender of the status quo. In other words, it is either your ally or enemy, and nothing in between.

The net contribution of technology to humanity has been astoundingly positive. Of all the eras of human history, now is arguably the greatest one to ever to bless the lives of so many multitudes. It is of course easy to lose sight of where humanity fits in the grand scheme of things, when the pettiness of politics, societal polarizations, war mongers, financial machinations, or specious social justice warriors dominates society's discourses. It is also easy to ignore the abhorrent injustices that remain etched into the fabric of many societies and cultures that have excluded themselves from the blessings of freedom and liberty. Nevertheless, the fact remains that the living standards that we take for granted exceed anything experienced by the kings, dictators, tyrants, and pharaohs of yore. We are heirs to phantasmagorical material, medical and liberal* successes, accrued over the past 500 years. Mankind, from times immemorial subject to the mercy and whims of Nature, painstakingly gained the upper hand over that period, graduating from subservient to master thanks to technology (consider the fact that we can construct tall buildings able to withstand earthquakes and tornadoes). Technology gradually unshackled people from the vagaries of Nature and Fate. A byproduct of this emancipation was the unleashing of wealth and freedom far and wide. This wealth empowered philosophers, empiricists, and scientists to change the world for the better. They banded together to construct a new humanist framework that put freedom first, which gave that framework purpose. This framework was empowered by the liberating endowment of technological breakthroughs (the printing press, calculus, the scientific method, chemistry, cotton clothing, electricity, telephones, the transistor, Boolean algebra, automobiles, planes, rockets, lasers, the internet, just to name a few). It spread its wings far and wide by the channeled power of

---

* The word "liberal" is chosen in the context of liberty and individual freedom, not in the divisive interpretation of the political class.

greed. Yes, greed. But not the greed of yore, spurred on to conquests by the whims of rulers; no, a new kind of greed, orchestrated into an operational concept called the firm. For there is no greater motivator than domesticated greed to stimulate innovations and quality across the widest possible market size to maximize gains. Greed, when pursued in the competitive arena, ultimately yields the form of our material existence today.* Adam Smith,† famed author of *The Wealth of Nations*, said it best:

> It is not from the benevolence of the butcher, the brewer, or the baker that we expect our dinner, but from their regard to their own interests.

Harnessing greed into a socially acceptable operating principle has not been easy—Emile Zola's *Germinal,* a novel about the horrors of the early days of the industrial revolution, is a chilling narrative about the dark side of greed. Greed has been, since the dawn of humanity, an individual prerogative shaped by the drive to survive. For thousands of years, greed was not so much the pursuit of personal gain at the expense of others (the quintessence of the zero-sum game) as it was a question of getting enough food to survive just another day, or getting the raw materials to fabricate one's tools, weapons, fire, and shelter. Indeed, the form stills exists today across all continents by millions of indigent and destitute souls. Greed emerged as a differentiated interest with the advent of widespread agriculture, sometime between 7000 and 10 000 years ago. The slow transition from hunting and gathering to domestication and cultivation created new social ecosystems in which large groups of people could be fed from grain surpluses. Specialization appeared on the

---

* Greed, evidently, can have a very dark side, much like Emperor Palpatine's dark side of the force... Things go dark really quickly when the level playing field of a properly governed competitive marketplace is forsaken by kleptocratic urges of dictatorships, nationalized industries, or crime syndicates. Without competitive pressures to act as checks upon corruptible power, markets no longer function while monopsonies thrive. Graft, seedy dealings, unrecoverable loans, and enrichment of the few rapidly take over. Concomitantly, everything positive associated with competitive configurations can erode: quality, reliability, transparency, efficiency, valunomy (see Chapter 3, and *Investment-Centric Project Management*, Chapter 2), and accountability get infected, like an incurable cancer. Hence the arrant importance of homogenous and rational regulatory constraints, placed upon the markets, by a conflict-free governing body, through industry standards, mandatory prescriptions, and legislated strictures.
† Adam Smith (1723–1790). Smith, a Scot, pioneered the study of political economy through the publication of two famous works: The Theory of Moral Sentiments (1759) and An Inquiry into the Nature and Causes of the Wealth of Nations (1776). The *Wealth of Nations* became to economics what Newton's *Principia Mathematica* was to physics. Source: Wikipedia. *The Wealth Of Nations*. Available at: en.wikipedia.org/wiki/The_Wealth_of_Nations. Last modified 27th February 2020.

scene, whereby individuals would focus on doing a limited set of tasks rather than operate as a survival generalist.* The invention of numismatic currency is generally agreed to have occurred in Mesopotamia around the middle of the third millenium BCE, with the silver shekel. The shekel manifested greed into a physical form.† Money is surely the greatest invention of mankind (fire and the wheel were discovered, not invented). Money enabled skill specialization to propagate across a society's strata, to produce food volumes in excess of survival, to build walled cities and roads and aqueduct and ships, and to create connections, via trade, between distant people and civilizations. It was also the promoter and enabler of wars, of conquests, of slavery, and tyranny. Like greed, money cannot be permitted to run wild in the hands of the few, lest injustice, indenture, and suffering of the great masses take hold. It requires a formal and legal framework to govern its dynamics. Notwithstanding, let us emphasize the *sine qua non* quality of money to the success of human endeavor. Without money, there is no travel, no discovery, no widespread healing, no invention, no technology, no modern life. It may have been mankind's willpower to overcome its many forms of dogmatic indentures, but it was money that transformed this willpower into action and creation. Without money, plans remain plans, aspirations remain aspirations. Action, inspiration, creation need money to bridge the chasm between concept and reality.

## *Who Is in Charge Here?*

Greed and money form a dynamic duo that underscores humanity's success in its timeless battle with nature for supremacy over destiny. With unbridled smugness, we Moderns can lay claim to victory over nature's whims. This victory is even—grudgingly—acknowledged in the suggestion that we live in a geological epoch of very own making: the *Anthropocene*, characterized by Mankind's hegemony over Nature (some good, some bad, admittedly). By conquering the power of energy, mankind has yoked Nature to its will and thrown off, in the process, earth's shackles upon the life, death and fates of societies. The vagaries of the Fates have been conquered by humans, if only when the political will decides to put on the mantle of dominance.

---

* Yuval Noah Harari's best-seller *Sapiens: A Brief History of Humankind* (2018, Harper Perennial; pp. 78–88) offers what is perhaps the most compelling description of the early days of agriculture and its consequences on the fate of humanity.
† Ibid, pages 173–188. In this case, Harari makes the case for money's crowning achievement: "money is the most universal and most efficient system of mutual trust ever devised" (p. 180).

In other words, it is Mankind that runs the show now, with Nature relegated to a supporting role. It is something that still seemed unfathomable a mere century ago.

---

### GEOLOGICAL TAXONOMY

The British geologist Arthur Holmes introduced the first geological time scale in 1913. He was also the first to estimate the age of the earth to be astronomically greater than religious dogma, at 1.6 billion years (today, the generally accepted age is 4.543 billion years). The modern form of the scale is divided into eons, eras, periods, and epochs, as shown in Table 1.1.

---

With the Anthropocene in full bloom, one could rejoice, justifiably, in the knowledge that mankind is at long last truly in charge of its own destiny— the good and the bad. Yet, beneath humanity's victory over earth's mercurial temperament, there lurks an ominous new realization. The long, arduous victory of Mankind over Nature appears suddenly fleeting. The digital construct is on its way to claim the hegemonic crown. There is a new ruler in town: data is the name and autonomy is the game. Whether in our daily lives, at the office, in the field, in the air, at the hospital, or in orbit, data are everywhere and requiring our obedience to their dictates. Data—pervasive, invasive, and omnipresent—have found their way into every conceivable human activity, to such an extent that societies have transformed into dictatorships of the machine over the individual. Doubtful? Then answer this simple question: is your smartphone within reach, as you read these lines? And if it rings or pings or sings, will you ignore it or set the book aside to answer its imperious interruption?

We may think that we lord over our devices, but the simple truth is that devices govern our waking and our sleeping moments, spurred on by the whims of software. Phones and tablets and laptops have become instant-gratification devices (aka *igeedee**).

---

* The neologism *igeedee* was originally coined by my wife Margaret in reference to the reflex that everyone in our household had acquired when an answer to a question was sought. Prior to smart-phones and tablets, an innocent question posed during a conversation (what is, say, the capital of Sri Lanka?) was answered in one of three ways: from someone's head, from a printed reference, or from turning on the computer—a tedious task at the time). But the handheld device changed the dynamics of things. At the push of a button, anyone could instantly find the answer from a screen held in the palm of the hand. Answers came instantly, without delays or doubts. The dopamine rush that came with this reassuring ability led to instant gratification. Hence the neologism *instant-gratification device*, or *igeedee* for short.

**Table 1.1   The Life and Times of Mother Earth Timelines Expressed in Million Years Ago (MYA)**

| *Era* | *Period* | *Epoch/Age* | *Start (MYA)* | *Events* |
|---|---|---|---|---|
| CENOZOIC | Quaternary | Holocene | Today | Ice age ends. Humans dominate |
| | | Pleistocene | 1.6 | Start of ice age. Earliest humans appear |
| | Tertiary | Pliocene | 5.3 | Human ancestors appear |
| | | Miocene | 23.7 | Green grass spreads far and wide |
| | | Oligocene | 36.6 | Dominion of the mammals |
| | | Eocene | 57.8 | Massive extension |
| | | Paleocene | 65.5 | Appearance of large mammals |
| MESOZOIC | Cretaceous | | 144 | K-T extinction. Appearance of flowers |
| | Jurassic | | 208 | Appearance of birds. Split of Pangaea |
| | Triassic | Age of dinosaurs | 245 | Appearance of dinosaurs |
| PALEOZOIC | Permian | Age of amphibians | 286 | Formation of Pangaea. P-T extinction |
| | Carboniferous | | 360 | Appearance of reptiles and cartilaginous fish |
| | Devonian | Age of fishes | 408 | Late Devonian extinction. Appearance of land animals and amphibians |
| | Silurian | | 438 | Appearance of land plants, jawed fish, and insects |
| | Ordovician | Age of invertebrates | 505 | O-S extinction. Appearance of vertebrates |
| | Cambrian | | 570 | Botomian extinction. Appearance of fungi. Dominion of trilobites |

(*Continued*)

**Table 1.1 (Continued)   The Life and Times of Mother Earth Timelines Expressed in Million Years Ago (MYA)**

| Era | Period | Epoch/Age | Start (MYA) | Events |
|---|---|---|---|---|
| PRECAMBRIAN | Proterozoic eon | | 2500 | Appearance of soft-bodied animals and multicellular life |
| | Archean eon | | 3800 | Appearance of unicellular life and photosynthesis |
| | Hadean eon | Priscoan period | 4600 | Formation of oceans and atmosphere |
| EARTH IS FORMED | | | | |

**GOTCHA**

Take a pause for a moment to observe the people in your vicinity. Count the people who are without any igeedee. Then, count how many are engaged in a real conversation (i.e., face to face) but with their igeedees on the table, near, or in their hands. How many will interrupt the conversation if the igeedee squeals for attention? How many people are sitting together but with their heads bowed down toward a screen? Finally, how many people are reading a printed text (other than business and contract paperwork)?

Even if you are in a meeting at the office, this experience will yield similar observations. People are addicted to their igeedees and are no longer exerting control over them. Their devices, and most likely yours as well, do. And this is just the tip of the point of the iceberg! At home or at the office, there will be a connection between you, the machines that surround you, and the inchoate ether that is the internet. All is connected, in real time, all the time, everywhere, for all times (Figure 1.1).

This transformation has been going on for decades, unbeknownst to everyone. It has progressed insidiously, first on the strength of the convenience experienced by users, then by the constancy of the connection. The allure of instant gratification from infinite access to information *at no cost* and the convenience of this access on the go, untethered to physical boundaries, have enfeoffed people (some might even say, unnervingly, enslaved) in everything that they do. People have so far gleefully accepted the Faustian bargain between their freedom and their convenience, in favor of the latter.

**Figure 1.1   Mexican Riviera vacation, igeedee style. Photo taken by the author in January 2018 at a resort on the Mexican Riviera.**

This embrace is universal and neither culture, religion, geography, or ethics will change the digital impetus, pecuniary conditions included. The tyrannical allure of the digital hegemon has spread beyond individuals. The internet of things, the cloud, and globalization have brought organizations, corporations, and governments in orbit around the digital planet. It would seem, whatever opinion the reader may have on the subject, that mankind has, willingly or not, chosen to be plugged in as a *modus operandi*. The task of living one's life unplugged has become Sisyphean or simply impossible. Data have been so thoroughly integrated into every facet of the modern (and not-so modern in some places) world that cutting the cord is unfathomable. By the same token, it has created a conflict between the motivations of people and the digital hegemon. The Symbiocene age will be marked by this conflict in the decades to come. As life's aspects become increasingly modulated by data and technology, the issue of technology as arbiter of who gets what and when, and who does not, will come to the fore. As *The Economist**\** noted, no technology is ideological in and out of itself, but takes on an ideological character through the actions of its proponents. The battle between

---

\* The challenger. *The Economist*, 17th March 2018. Available at: www.economist.com/briefing/2018/03/15/the-challenger.

open-source and proprietary software is a glaring example (benefit of the commons against benefits of shareholders). It acquires an ominous dread when its capabilities enable tyrannical ideologies of the kind promulgated by China's social credit system.

Through this choice, people and organizations have permanently altered the implicit compact that had existed since the dawn of time between supply and demand, between need and want, or between benefit and greed. Before Tim Berners-Lee changed the world forever,* this compact was rooted in a tacit tension between the priorities of the *demander* and those of the *supplier*. Then, the *demander* was in the driver's seat: she could not be unwillingly forced to buy the *supplier's* offering. In the *Symbiocene Age*, at least for the time being, this tension is no longer ruled by the *demander*; it has swung over to the *supplier*. Not convinced yet? Try avoiding upgrades to the operating system or the apps that are on your computer or your *igeedee*. Or try to download a free app on your phone without agreeing to allow the app to have access to your pictures, your contacts, your schedule, and your browsing history.† Even software, as a product category, has never been bought by anyone: it is used under a license governed by excruciating legal terms that boils down to this: this software makes no claim that it will work for the purposes for which it was bought. You can buy an airplane but you cannot own the software that makes it fly. The *supplier* has the upper hand, and in fact the monopsony over the supply-and-demand compact. Data impart a permanent dependency (or addiction?) upon their users—the *demanders*, who are no longer driving the bargain. Data do.

---

* Sir Tim Berners-Lee (1955–) is best known as the inventor of the World Wide Web, which gave the world the internet. English engineer, computer scientist, and professor of computer science at Oxford University, Berners-Lee's brilliant insight was to conceive an information management system, erected upon the hypertext transfer protocol (HTTP), to establish digital communications between a computer and the rest of the internet. The advent of the World Wide Web, in 1984, ushered in the first primitive instances of what this book calls the age of the Symbiocene.

† My smartphone runs the Android operating system. I once tried to install on it the free version of Microsoft Excel. This is the message, word for word, that popped up on the screen: *Allow Excel to access photos, media, and files on your device?* I asked myself why on earth does Excel need to access my photos in order for me to use the damn app? The answer is, of course, the nefarious agenda of the supplier, to monetize *my* data without paying nothing for it. The reason, in any case, is arrantly irrelevant. No app requires access to my photos and media files to run. Except that I, as a *demander*, have no say or choice in the matter. Either I accept this coercive condition, or I do not get to use the app. As a result, Microsoft apps have forever been banned from my phone, as well as ALL other apps that insist on this egregious condition. Sadly, I suspect that my stance is in the minuscule minority.

## *To Be or Not to Be, the Eternal Question*

The final reason for getting onboard the digital bandwagon has to do with the egregious effects of the status quo. The Symbiocene age is no country for old ways. It does not matter whether a business has been around for a century, a decade, or a year. The old ways of running a business cannot compete in this new arena. Neither can staid business models comfortably ensconced behind the illusion of manageability. But old ways do not mean old age. As a matter of fact, old age is a benefit to the enterprising transformer, because of the battle scars of the past. Experience has the advantage of knowing that change is neither simple, smooth, or efficient. Experience knows that one must get into this game with eyes wide open, to be prepared for things to go awry. Experience knows that plans never survive contact with reality. Experience, above all, knows that success comes from keeping an open mind when challenging the specious intimations of the status quo.

*The age is not a country for young ones either.* It is easy to be lulled into the belief that the younger generations, raised on igeedees while still in diapers, are better suited to the role of front-line warriors in any transformation engagement. Proficiency with texting, with social media, and with mobile transactions are essentially *useless* in a binary firm. The digital construct does not simplify the inner workings of a business; it exacerbates their complexity. Running a binary firm, anchored to the data hegemon, requires a mastery of functional corporate processes that has nothing to do with digital familiarity and everything to do with relational complexity. Such mastery is not the domain of the MBA either. It is, above all, in the realm of the business professional who understands how the business makes and spends money. This is the third digital paradox of the Symbiocene age:

> Paradox 3: Digital literacy and social media numeracy are of tertiary importance to the successful binary firm manager. The secondary importance is the underlying mechanics and mechanisms implicated in the costs and profits of the business. The primary importance is the emotional intelligence around the interactions of people and digital machines.

# Why This Book?

## *The Point*

This book deals with the design of a business or an organization against an ever-changing nature of the environment in which it operates. It is a guide to the reader on the journey to become a binary firm, and for grappling with the ramifications of managing the baffling complexity that accompanies technological wonders. It provides the reader with strategies to deploy capital investments rationally, with a view to account for future trends that may threaten the investment.

> The world beyond 2025 will be nothing like the world of business today. Nobody can anticipate where technology will truly be a mere five years out. Hence, the only viable survival strategy for businesses large and small is to design their organizations to: 1) be able to respond immediately to whatever disruption comes their way; and 2) be dynamically malleable to change their modus operandi on the fly so that these disruptions become wealth-generating opportunities.

The point of this book is to equip executives and managers with practical tools and strategies to transform their organizations into dynamically malleable entities, controlled by *anticipatory management*, and operated in such a way that it profits in real time from whatever technological disruption comes their way, *without breaking the bank or the business model*. In short, a business model that is itself inherently disruptive and designed to assimilate the upside of technology without the downside of hits to the bottom line.

Lastly, this book makes no attempt to predict where technology is heading, nor the shape of the world beyond 2030. Yogi Berra's famous quote about "you can observe a lot by watching" serves as a guiding principle, complemented by another one of his: "it's tough to make predictions, especially about the future". After all, nobody in 2008 envisioned the astonishing success of the electric car in 2018. There are simply too many variables in play to achieve any meaningful reliability in predictions, especially about the evolution or revolution in technology and society. What is possible, however, is to extract the lessons from natural selection and apply them to a *predictive* approach to anticipatory management. We do not know what is coming down the pike. What we do know is that the present is already on its way out. Better equip yourself with the means of thriving from the future, instead of perpetually fighting it for your survival.

# Bibliography

Ampere, Andé-Marie. *Essai sur la philosophie des sciences*. University of Michigan Library, 2009, 302 pages.

Babbage, Charles. *On the Economy of Machinery and Manufacturers* (original 1832). Reprinted by Cambridge University Press, Cambridge UK, 2009, 320 pages.

Bergreen, Laurence. *Columbus – The Four Voyages*. Viking Press, New York, NY, 2011, 422 pages.

Blanning, Tim. *The Pursuit of Glory – Europe 1648-1815*. Penguin Books Ltd, London, 2007, 708 pages.

Boole, George. *An Investigation into the Laws of Thought*. Palala Press, 2015, 438 pages.

Cann, Geoffrey; Goydan, Rachael. *Bits, Bytes, and Barrels: The Digital Transformation of Oil and Gas*. MadCann Press, 2019, 209 pages.

Diamond, Jared. *Guns, Germs and Steel – The Fates of Societies*. W.W. Norton and Company, New York, NY, 1997, 480 pages.

Engels, Frederick. *The Condition of the Working Class in England* (original 1845). Reprinted by Pinnacle Press, 2017, 344 pages.

Finlay, Steven. *Artificial Intelligence and Machine Learning for Business: A No-Nonsense Guide to Data Driven Technologies*. Relativistic Books, London, UK, 2017, 150 pages.

Freeman, Joshua B. *Behemoth – A History of the Factory and the Making of the Modern World*. W.W. Norton & Company, New York, NY, 2018, 427 pages.

Harari, Yuval Noahi. *Sapiens: A Brief History of Humankind*. Harper Perennial, 2018, 464 pages.

Harris, Karen; Kinson, Austin; Schwedel, Andrew. *Labor 2030: The Collision of Demographics, Automation and Inequality*. http://bain.com/publications/articles/labor-2030-the-collision-of-demographics-automation-and-inequality.aspx.

Higginson, Matt; Nadeau, Marie-Claude; Rajgopal, Kausik. *Blockchain's Occam Problem*. McKinsey & Company, January 2019. https://www.mckinsey.com/industries/financial-services/our-insights/blockchains-occam-problem.

Holmes, Arthur. *The Age of the Earth*. Forgotten Books, 2017, 222 pages.

ISO standard 15489-1. Information and *Documentation - Records Management, Part 1 – General*.

ISO standard 15489-2. Information and Documentation - *Records Management, Part 2 – Guidelines*.

Keays, Steven J. *Investment-Centric Project Management: Advanced Strategies for Developing and Executing Successful Capital Projects*. J. Ross Publishing, Plantation, FL, 2017, 419 pages.

Keays, Steven J. *Investment-Centric Innovation Project Management: Winning the New Product Development Game*. J. Ross Publishing, Plantation, FL, 2018, 309 pages.

Kelsey, Todd. *Surfing the Tsunami – An Introduction to Artificial Intelligence and Options for Responding*. Todd Kelsey Publisher, 2018, 188 pages.

Kranz, Maciej. *Building the Internet of Things – Implement New Business Models, Disrupt Competitors, Transform Your Industry*. John Wiley & Sons, Hoboken, NJ, 2016, 272 pages.

Mar, Bernard. *Data Strategy: How to Profit from a World of Big Data, Analytics and the Internet of Things.* Kogan Page Limited, London, 2017, 200 pages.

McLuhan, Marshall. *Understanding Media: The Extensions of Man.* MIT Press, 1994.

Morison, Samuel Eliot. *Admiral of the Ocean Sea: The Life of Christopher Columbus.* Atlantic/Little, Brown, Boston, MA, 1942. Reissued by the Morison Press, 2007.

Rifkin, Jeremy. *The Third Industrial Revolution: How Lateral Power is Transforming Energy, the Economy, and the World.* St-Palgrave MacMillan Publishers, New York, NY, 2011, 304 pages.

Rogers, David L. *The Digital Transformation Playbook.* Columbia Business School Publishing, New York, NY, 2016, 278 pages.

Rossman, John. *The Amazon Way on IoT: 10 Principles for Every Leader from the World's Leading Internet of Things Strategies.* Clyde Hill Publishing, Washington, DC, 2016, 168 pages.

Russell, Stuart. *Artificial Intelligence – A Modern Approach.* Pearson Education Limited, London, 2015, 1164 pages.

Sacolick, Isaac. *Driving Digital – The Leader's Guide to Business Transformation Through Technology.* American Management Association, 2017, 283 pages.

Schwab, Klaus. *The Fourth Industrial Revolution.* Crown Business, New York, NY, 2017, 192 pages.

Sills, Franklyn. *Foundations in Craniosacral Biodynamics, Volume One: The Breath of Life and Fundamental Skills.* North Atlantic Books, Berkeley, CA, 2016, 424 pages.

Sloman, Steven; Fernbach, Philip. *The Knowledge Illusion: Why We Never Thing Alone.* Riverhead Books, New York, NY, 2017, 304 pages.

Smith, Adam. *An Inquiry into the Nature and Causes of the Wealth of Nations.* The University of Chicago Press, 1976, 1152 pages.

Smith, Adam. *The Theory of Moral Sentiments.* Digireads.com Publishing, 2010, 238 pages.

Tapscott, Don; Tapscott, Alex. *Blockchain Revolution: How the Technology Behind Bitcoin Is Changing Money, Business, and the World.* Portfolio/Penguin, 2016, 432 pages.

Vigna, Paul; Casey, Michael J. *The Age of Cryptocurrency: How Bitcoin and the Blockchain Are Challenging the Global Economic Order.* Picador, New York, NY, 384 pages.

Westerman, George; Bonnet, Didier; McAfee, Andrew. *Leading Digital – Turning Technology into Business Transformation.* Harvard Business School Publishing, 2014, 256 pages.

Wiener, Norbert. *Cybernetics - 2nd Edition: Or Control and Communication in the Animal and the Machine.* The MIT Press, Boston, MA, 1965, 212 pages.

Windpassinger, Nicholas. *Internet of Things: Digitize or Die: Transform Your Organization: Embrace the Digital Evolution: Rise above the Competition.* IoT Hub Publishers, New York, NY, 2017, 284 pages.

Zola, Emile. *Germinal.* Penguin Books, London, England, 2004, 592 pages.

## Chapter 2

# The Symbiocene Age

*Evolution is like walking on a rolling barrel. The walker isn't so much interested in where the barrel is going as he is in keeping on top of it.*

**Robert Frost, *The Letters of Robert Frost to Louis Untermeyer***

The fourth industrial revolution will be technology driven, like its predecessors. However, the ramifications go beyond another case of the new superseding the old. They are harbingers of more profound transformation of the relationships between people, between people and machines, and between machines. The upheaval of the contemporary order arising from these catalytic changes is forging the new age of the Symbiocene. For any digital transformation, success entails a grasp of the history that led us to this point.

## A Brief History of Now

### *Weaving a Tapestry*

Technology has played a seminal role in the chaotic evolution of human societies. The use of fire to cook enabled a rapid growth in the brain size of early hominids.* It also powered the very first technological revolution—the Neolithic—with the advent of metal tools and farming. Metal tools turned into weapons that made territorial conquests possible. Wars and agriculture led to domiciled civilizations, political systems, and labor differentiations. A technological thread is thus woven into the very fabric of humanity. But technology remained subservient to its human masters, who deployed it in all its myriad forms to their pursuits of power. Power was always the overlord, the maker of all things. Power was, and remains to this day, a deeply ingrained evolutionary instinct toward self-preservation. Survival of the fittest and elimination of the weakest went hand in hand.† Thus was born the timeless partnership between technology as enabler and servant of power (which begat the mother of all technological disruption: money).

By virtue of its station, power became the catalyst for all manners of evolutions and revolutions until this day. Europe emerged from the rubbles of the French Revolution and the Napoleonic Age with the first industrial revolution in tow. Its primitive technology began to play an immediate transformative role in society to an extent never seen before. Large cotton mills appeared on the horizon, hitherto bereft of crowded estates, on a scale unimaginable to previous generations. Charles Babbage, the famed mathematician, was among the first to comment on this "gigantism" trend, driven by the ability to achieve great production volumes with the introduction of machinery housed under one roof.‡ Factories wrought wrenching changes to the nature of labor and working conditions. Whereas, since

---

* The modern human brain consumes about one-fifth of the body's energy, by far superior to any other creature in the animal kingdom. The consensus among anthropologists posits a vegetarian diet for early hominids, which could not have supported such an energy expense by their brains. The extra energy had to come from somewhere and meat is the most plausible source. The anatomical evidence points to such an adaption about 1.8 million years ago, just as brain size begins to increase. Eating raw meat, however, is itself energy-consuming (as any satiated lion will show…). Adding fire to the mix completely alters the energy balance in favor of the eater: cooking frees up energy from the ingested foods. This is the insight of Wrangham and his Harvard colleague Rachel Carmody. What matters, they say, is not just how many calories you can put into your mouth, but what happens to the food once it gets there. Cooking is akin to a free lunch! Incidentally, it is a simple leap to infer that cooking became a daily event that drew the clan together, which might have spawned language in the process.

† The idea that natural selection is in fact powered more by the elimination of the weakest than the survival of the fittest is explored in Chapter 12 of *Investment-Centric Innovation Project Management*.

‡ Charles Babbage (1791–1871) was a philosopher, inventor, and mechanical engineer, but mainly famous as a mathematician and progenitor of the idea of a programmable computer. It is fascinating to note that all the essential concepts and ideas of modern computers were discussed in his mechanical invention: the analytical engine. Babbage's commentary on the effects of the factory upon society appeared in his 1832 book *On the Economy of Machinery and Manufacturers*. See also Joshua B. Freeman's *Behemoth: A History of the Factory and the Making of the Modern World* (2018, W.W. Norton & Company).

times immemorial, people had eked out a living in relative isolation as artisans, farmers, or in servitude, factories agglomerated swarms of workers from all walks of life and clustered them to tend to machinery in appalling conditions.* As grim as the impact on people was, it paled in comparison to the impact on villages and cities. Hundreds or thousands of workers employed *inside* a facility created Sisyphean challenges in transportation, housing, sanitation, agriculture, and feeding for which the requisite infrastructures did not exist. Hard currency ran into acute shortages when everyone was getting paid rather than transacting their daily needs on the basis of barter. Civil order and worker discipline began to break down in the face of migratory people accustomed to living among the few but forced to commune with hordes of strangers.† *Luddism*, the reactionary movement that appeared on the scene in the 1830s and 1840s, was a desperate cry by millions of suffering souls who were protesting their desultory factory existence. The thrust of it would eventually morph into the union movement of the early twentieth century.

## *Change of the Guard*

The ineluctable march of technology was sustained by those who wielded its power. Progress, in the name of modernization but at the silent price of endless toils, continued unabated, like a giant glacier grinding the land underfoot from its advance. Landscapes were forever and irreversibly changed, often leaving destruction and destitution far and wide. Cultures and mores of yore were no bulwark against the technological juggernaut, with power and greed at the helm. The dynamic is still in play today, as we observe the relentless concentration of economic heft in the hands of digital corporate behemoths. On the other hand, there are signs that technology is gaining strength of its own and showing signs of challenging its state of servitude to its corporate masters. Technology is in fact proliferating *on its own accord*, without the staying hand of corporate strategists or political tyrants. Evidently, technology is borne in the bowels of manufacturers. But its explosive growth across virtually all facets of human endeavors has

---

* Engels, for instance, noted in his book *The Condition of the Working Class in England* (Originally published 1845; Reprinted: 2017, Pinnacle Press; pp. 174, 199) that the tedious and repetitive act of workers in tending to machines was tantamount to torture sustained in a never-ending cycle by the machines themselves.

† See Joshua B. Freeman's *Behemoth: A History of the Factory and the Making of the Modern World* (2018, W.W. Norton & Company), Chapter 1.

given technology the most potent network effect possible.* It is this network effect that is powering the profound relationship changes mentioned earlier. Technology, in other words, has achieved a critical mass *of its own*, impervious to the whims of the old power guards. The changes spawned by the ever-expanding reach of technology are leading the transformative consequences with which we must now deal. Welcome to the Symbiocene age.

That is why this age is different from its predecessors. It is not simply ushering in yet another wholesale replacement of old with new tech. *In fine*, our prime concern has nothing to do with choosing which fangled software to buy, what data servers to install, or what web apps to develop. Our concern is with the effects of technology upon the relationships that make the business world go around. The human experience of relating to one's world is in fact being profoundly affected. As Franklyn Sills put it: "Our relationships are the ground of our experience. It is from our relationships that we mold our sense of the world and our place within it. Thus the work, in a powerful way, can only bring us back to ourselves".†

## Impacts of the Age

### *The Circle of Ignorance*

In a historical sense, the Symbiocene age marks a full circle to the time before the first revolution. In the late eighteenth century, ignorance was the dominant state of humans. Most people had no clue how the world worked. Religious fantasies and magical beliefs played the role of theories to explain why things were. The vast array of mathematical and scientific discoveries was in full swing but stood out mainly because it sat in the middle of a dark ocean of cluelessness. Knowledge began to spread beyond scientific circles when James Watt, the system engineering genius, introduced the first productive steam engine in the 1760s, with the power and reliability (always under-rated yet never over-appreciated) to underwrite the rise of the

---

* The network effect is gives its holder a winner-takes-all advantage, through which the value of a product or service increases with increasing numbers of users of that product or service. That is why Google, for example, is the undisputed king of online searches.
† Sills' context about the word "work" is derived from his expertise in the craniosacral field. Sills speaks of the biodynamic craniosacral work as the creation of a relationship between the biodynamic practitioner and her client that is rooted in a direct and safe human connection. This stands in opposition to the modern habit of automatically injecting technology as mediator between one's action and intent. See *Craniosacral Biodynamics* by Franklyn Sills (2016, North Atlantic Books; p. 4).

machine. This was a complicated machine, in a modern sense of the world, one that required practical knowledge in manufacturing, metallurgy, chemistry, mathematics, and logistics, from the many people involved in its economic ecosystem. Knowledge kept the machines running. And the machines found the "killer app": cotton underwear.*

### SYSTEM ENGINEERING, RENAISSANCE STYLE

Contrary to popular belief, James Watt (1736–1819) did not invent the steam engine. That honor belongs to Hero of Alexandria (circa 70 CE), with his spinning sphere *aeolipile*. The ingenious device was not, as far as it is known, used as a power machine. The first machine to be used to produce mechanical power was invented by Thomas Newcomen in 1712. But low efficiency, low reliability, and bulk did not make Newcomen's invention the disruptive technology that it would become in the hands of Watt. Watt's brilliance was to improve the design by adapting a number of existing and new solutions to solve the pressing issue of the enormous waste of energy incurred in cooling the power cylinder. He added a separate condenser, a centrifugal governor, piston double-action, a steam indicator, and changed the configuration to generate circular motion. The result was an engine that was 500% more efficient than Newcomen's baby.

In this respect, Watt followed in the hallowed footsteps of another genial system engineer, Johannes Gutenberg (c. 1400–1468). The printing press was already in widespread use in Europe, based on a Chinese design. The system, however, was laborious, slow, and prone to gumming up. Gutenberg improved on the Chinese design with the introduction of metal printing matrices, used as movable types. The metal matrices were molded by hand, which made it possible to rapidly produce them in large quantities.

---

* The textile industry was the first large-scale industrial production to take advantage of the steam engine (the mining industry was a close second in a tie with rail transportation). Textile production was, up till then, the epitome of manual labor. Mechanized weaving powered by steam engines changed all that, as is well known. What is less known is why there was such a sudden increase in textiles. After all, it is not like there was a population explosion. The answer, amazingly, was the humble cotton underwear. The cotton industry saw an explosion in production volumes in the 18th and 19th century from the abhorrent slavery practiced in the New World on vast cotton plantations. The softness and comfort of cotton went beyond anything that the great plebian masses had seen. Cotton underwear became the "killer app" of its day. Suddenly, even the poor could afford the wonder of the ages. This led to a global market for all things cotton. The production volumes required to satisfy this demand required the industrialization of textile production. See Tim Blanning's *The Pursuit of Glory – Europe 1648–1815* (2007, Penguin Books Ltd; p. 135).

The introduction of the movable type proved seminal to the history of mankind. It yielded dramatic production cost reductions, which made it possible to disseminate printed works far and wide across the continents.

The second industrial revolution, early in the 20th century, gained impetus from Henry Ford's introduction of mass production methods. These methods rapidly found new homes wherever things were manufactured; their spread brought with it the idea of the *firm*, which took over the levers of the economy across the nascent Western order. The conduct of business, which had hitherto been a bipolar order between domineering oligarchs and family owned small businesses, splintered into a continuum of business sizes unheard of in terms of range. Ford's production line schema also led to the invention of management as a formal science taught by universities and colleges (birthing, along the way, the MBA degree). Once again, the stock of knowledge required to run these firms expanded exponentially in contents and practitioners.

The arrival of the internet in the 1980s marked the dawn of the third industrial revolution. It laid down the foundation of the world we live in, where people and organizations connect to each other from anywhere to elsewhere. Globalization happened. China was reborn. And the creation and dissemination of knowledge went beyond exponential. Then, something unexpected happened along the way: the ignorance paradox showed up. On the one hand, we have a staggering amount of information being generated and transacted every second of every day. This production is a flood, a deluge, a tsunami even. On the other hand, it is utterly beyond the grasp of the human mind. That is the ignorance paradox (and the fourth of the Symbiocene age):

> Paradox 4: The proliferation of information causes a generalized dilapidation of individual knowledge.

With the fourth industrial revolution running at full tilt, knowledge has now escaped the human orbit to be captured by intelligent machines, intelligent algorithms, and self-learning software. Machines are generating knowledge *without any human guidance.* Sophisticated neural networks, for example, are able to arrive at the right conclusions without anybody understanding how they did it. Not only is knowledge out of human reach, it is becoming unreachable by the mind.

That is the meaning beneath the statement that history has come full circle. Three hundred years ago, most humans were utterly clueless about the ways of the world. Knowledge existed on minuscule islands populated by very small tribes of intellectual explorers. Nowadays, most humans (especially those in so-called advanced economies) go about their daily lives without an iota of understanding of the how the world works. Most humans float in a vast sea of ignorance peppered with innumerable islands of acute knowledge and expertise equipped with arduous docking facilities. To paraphrase Churchill, never in the annals of time has so much knowledge been available to so many and understood by so few. To make matters worse, the awesome connectivity of the digital hegemon has lured unsuspecting masses of people into a state of perpetual dependency upon it. Convenience coupled with instant gratification (the foundation of all igeedee devices) becomes, to the digital master, the killer app for ensnaring people in its web. People have willingly given up their independence of thought and outsourced it to the web. How many readers can drive to an unknown address *without* relying on GPS? How many readers will borrow or buy a book to explore in greater details a subject browsed from Wikipedia? How many readers will counter check any "fact" presented in some online blog about whatever topic? Who can recall the numbers of their children's smartphones? Very few indeed. The result? Digital information's veracity is taken for granted, even in the commercial and business realms. That is a dangerous presumption to embrace as a *modus operandi* under the best of circumstances. It leads to a nefarious effect of the digital hegemony:

> Convenience is the stepping stone to subservience, in the Orwellian sense.

Incidentally, the dangers of blind faith in digital information must be front and center to any digital management strategy and are treated as such in this book.

## The Rise of the Datum

The accretion of ignorance appears to be an unavoidable consequence of the Symbiocene age: the more we click, the less we know. Furthermore, the more we click, the more autonomous data becomes, which is part of the reason why technology has reached critical mass. We are witness to a startling development in the machine realm as well: data are becoming *functionally sentient* and simultaneously independent of human action (more on this later in this chapter). Humans do not handle ignorance and disfranchisement well. History reveals that humans tend to cope with these fears through mechanisms that are nearly always self-destructive: superstitions, tyrannical structures, misogyny, xenophobia, misanthropy, luddism. But history is our best teacher for avoiding those behaviors.

The Symbiocene age is a *thing* because of one word: *data*. Data now rules the kingdom. It is the progenitor of all trends. Computers, automation, and robots breathe data in and out to exist just as organics breathe to survive. The gargantuan size of this datum production happens invisibly, without smell or noise (save for the cooling fans sitting at the back of the server racks). Data transit across the internet, live in giant data servers, and span the globe through the Cloud. The internet, at once enabled and enabling, omnipotent and ubiquitous, has relinquished its primacy over the world order. It has devolved into the plumbing system through which data and information transit. The new claimant to the technological throne is a numeric trio: data, information, and software, the modern equivalent of the Roman triumvirate* lording it over a global empire. Their invisible relationships are evolving outside of physical time, spreading their consequences imperviously to human or corporate volitions. Out of these relationships is emerging a symbiotic phenomenon that is greater than the sum of its parts: *functional sentience*.

## Functional Sentience

During the entirety of the third industrial revolution, the data that have been created, streamed, and transacted by software, machines, and sensors have been inorganic (inert as stones, in other words). They were given passive binary forms but denied any collective ability to "act" out of their own volition. Some sort of algorithm (numeric or analog) was required to transform them from inputs to outputs and vice versa. These algorithms were also inorganic as well; relying on precise codifications of the transformation procedure, without any possibility of self-modification of those coded instructions. Algorithms were embedded into closed-form software that could execute magnificent computations *as long as these computations conformed to the absolute constraints of the coding sequences.* Software, in other words, were astonishingly beautiful dumb machines limited by the adage "garbage in, garbage out".

---

* The first Roman triumvirate came about when Caesar, Pompey, and Crassus agreed to an alliance, between 59 and 53 BC, to rule what would effectively become the last Roman republic. You could say that the alliance did not end well…

The advent of the fourth industrial revolution exerted a dramatic transformation in the inorganic nature of data, information, and software. The transformation emerged from the fusion of the triumvirate into "organic" systems that were devoid of immutable coded instructions (artificial intelligence, neural networks, and machine learning). These systems are organic in the sense that they are self-controlled, self-correcting, self-replicating, and adaptable to external conditions. They can take garbage in and still turn it into useful outputs. Concomitantly, the emergence of widely distributed data networks and distributed ledged technology (DLT)—aka *blockchain*—enabled the deep integration of innumerable datum-generating sourced into cohesive transaction networks. Complex sensor-based networks of machines and operations have arrived under the banner of the Internet of Things (IoT) and the Industrial Internet of Things (IIoT). When the organic software systems are married to the IoT or the IIoT or both (an instance of valuable polygamy), the outcome is yet another emergent behavior: *functional sentience.*

The Chambers Dictionary* defines "sentience" as the capacity for sensation or feeling; for consciousness and awareness. Chambers also defines "functional" as what is designed for efficiency rather than decorativeness; plain rather than elaborate; in working order or operational. Together, the words yield an operational concept of the new digital age:

> Functional sentience is an emergent behavior of data-driven systems whereby a system is aware of the external environment and its intrinsic performance and is able to autonomously adapt itself to preserve its pre-defined performance mandate.

Functional sentience is characterized by five primordial characteristics:

1. The ability to assess, in real time, the state of the inputs supplied to it, the outputs supplied by it, and the correctness of its interactions with the external environment.
2. The ability to adapt, in real time, its internal data treatment processes, in a feedback loop with the inputs, when outputs are found delinquent.
3. The ability to perform transactions autonomously between its internal processes and the external data sources and data clients.
4. The ability to gauge its performance and detect functional and physical faults, data degeneracy, assessment biases, and calibration drifts, and generate appropriate alerts and responses to human overseers.

---

* *The Chambers Dictionary, 13th Edition.* John Murray Press, London, UK, 2014.

5. The ability to operate within its design operational envelope autonomously, without constant or continuous human interventions.

*Functional sentience* captures the essence of the transformative character of the Symbiocene age. Vast, complex systems are already operating in various places over the globe with some or all of the five primordial characteristics active. They are still primitive in some of these abilities but are nonetheless functionally sentient in practice.* The bottom line is this: functional sentience is upon us and will only explode in capability and extent with the passage of time. Its correlation to our discussion on digital management transformation is straightforward: how does an organization exploit the potency of functionally sentient systems without losing control of management's decision relationship with the overall business operation?

## A New Continent

### *Tectonic Shifts*

*Functional sentience* is a direct result of the convergence of the physical, digital, and biological worlds described by Klaus Schwab in *The Fourth Industrial Revolution.*† This convergence is birthing new operating concepts

---

* The deployment of smart energy grids is an example of an early functionally sentient system. Such grids connect power producers and users into self-balancing power distribution networks. On these networks, a producer can become a user and vice versa, for instance, when a solar panel operator "feeds" power to the grid in daylight or at night from a battery farm, and receives power from the grid when the weather is cloudy or batteries are depleted. Scaling this example to hundreds or thousands of individual participants creates enormous complexities in terms of "balancing" the power flowing to and from the network. Balancing is accomplished in real time by autonomous monitoring systems, and in tandem with billing systems that can adjust the power rates on the fly as a function of immediate demand. The autonomy would not be possible as a centrally directed management scheme. It is possible through a massive devolution of decision powers to the individual components of the grid, like sensors analyzing consumption data *in situ*; electric smart meters deciding when the direction of energy flow must change; meters billing the network owners directly; battery storage installations requesting power inflows by themselves below a given threshold; the same installations adjusting the thresholds as a function of recent weather and demand trends; system monitoring sensors alerting the local owner of a faulty component; and power monitoring algorithms detecting hacking or tampering attempts on a grid node, and activating an isolation protocol to contain the attack.

† Schwab, Klaus. *The Fourth Industrial Revolution.* Crown Business, New York, NY, 2017, 192 pages.

deployed or employed by all matters of disciplines, economies, industries, and governments. It is occurring in real time, on a global scale, with untold complexity and raging under currents assailing the status quo everywhere. These currents are a manifestation of deeper tectonic impacts, akin to Earth's geological tectonic plates.* On Earth, tectonic plates carry all land masses on their back. About 335 million years ago, the plates formed a single land mass, a kind of supercontinent named *Pangaea*, which began to break up about 175 million years ago with the separation of the plates. The plates are continuously moving relative to each other or above one another (a process called subduction). When they collide, the likes of Mount Everest are the result. When subduction snaps, earthquakes and tsunamis ensue. In other words, the motion of tectonic plates dictate how life lives on this planet of ours.

The plate tectonics provide us with a vivid analogy to what is happening now. The world today is the geographic heir to the breakup of Pangaea into the familiar continents. Humanity is in turn the inheritor of the great dislocations and shearing forces unleashed by the first three industrial revolutions. These forces created distinct economic territories (first, second, and third worlds), geo-political plates (the West, the Near East, the Far East, the Middle East), and industrial ranges (OECD, BRICS, OPEC, Silk Road, etc.). Knowledge islands have sprouted everywhere in a joyful syncopate chaos nurtured by research funding, intellectual property frameworks, government priorities, and investment greed. Life is lived at different paces as a function of one's station within each of these coterminous delineations. Businesses also thrive or choke depending on the constraints of their geographic locations and technological maturity.

There are of course several different ways to divide *Homo Pangaea*, about which we make two core observations: 1) the human world is dividable into distinct morphological groups; and 2) these groups are connected by sinews of technology. Groups will take any number of identities according to the underlying values buttressing them. Some will be violently distinct (theocracy vs democracy for instance) while other will form a coterminous continuum of shared values (European vs American politics, China vs OECD). All human groups, regardless of their degree of xenophobia, will

---

* The discovery and acceptance of the theory of tectonic plates is an illuminating example of the belligerent reaction of the established *status quo* to a seminal innovation. Plate tectonics were introduced by Alfred Wegener in 1912 under cover of a new theory of continental drift. The clash of scientific dogma and necessary advancement in knowledge would span five decades before the futility of the status quo would be finally accepted.

maintain relationships with others—no matter how tenuous they could be—which gets us to the crux of the matter:

> Relationships are the conduits through which ideas and technologies migrate and evolve across space and time.

In this single sentence lies the genesis of our nascent digital world. Technologies will spread ever farther and wider as long as people maintain exchanges among themselves. This discovery was, to this author's knowledge, first discussed by Jared Diamond in his 1997 bestseller *Guns, Germs, and Steel.*\* It explained why, for instance, the people of Mesopotamia, who were first to domesticate plants and husband animals, were also the first to invent writing, set up governments, invent new technologies, and acquire natural defenses against bugs and germs. It is a simple step to take the conclusion further: relationships are inevitably changed by their intimate contact with changing technologies. In other words:

> Human relationships and technology exist symbiotically, in ever increasing intimacy as time goes by.

The reader should note that the above statement is quiet on the merits of this symbiosis, which can be either positive or negative, creative or destructive, according to what a human decides to do (or not do) with technology. Technology exists beyond good and evil (to borrow from Friedrich Nietzsche); it merely *is*. It becomes good or evil through human intent (like guns, germs, and steel, incidentally). Regardless, it is the fact that technology and relationships are intertwined that matters.

## Digital Magma

We are now in a position to ask ourselves what are the technological threads that run through the *human Pangaea*. These threads do more than stitch the plates together: they are actually pulling them back together. In effect, they are causing a *reversal* in the divergence of Pangaea: we are witnessing

---

\* Jared Diamond. *Guns, Germs, and Steel: The Fates of Human Societies* (1997, W.W. Norton and Company). Diamond's thesis on the origins of the hegemony of Western civilization provides an eminently credible explanation on why different peoples followed different evolutionary courses according to the particulars of their immediate environment, and not at all because of putative biological differences.

for the first time in history a trend toward unification. The great digital convergence runs through the internet, the mother of all technological threads, the virtual tapestry upon which are inscribed the patterns of transformation. The tapestry is woven from five main threads offered in several hues: the power chip, artificial intelligence, distributed ledgers, the energy lattice, and the internet of things. The glue that holds everything together is data—which is a class of its own and discussed in the next chapter.

> Data is judge, jury, and executioner of any business seeking to thrive in the Symbiocene age. It is input and output, fuel and power, cost and profit simultaneously.

# The Power Chip

## *The Symbiocene's Power Plant*

The relentless advances in chip design has powered the third industrial revolution and doing the same for the Symbiocene. Computer chips are so ubiquitous as to be invisible and forgotten. Nevertheless, the miracles of computations that they perform are the reason why we have cell phones, GPS, self-driving cars, MRI machines, cancer drugs, high density lithium batteries, robots, and efficient air filters. There is no aspect of the physical world that is not handled by one chip or another. With millions of transistors packed into the size of a thumb tack, the humble microchip has truly taken on the mantle of engine that powers our world. Sophisticated algorithms can be imbedded into them to operate autonomously, making them deployable in infinite applications from the mundane to the extreme. Batteries and lightweight rotors may keep drones airborne, but it is the microchips they carry that make them useful and commercially viable.

> The entirety of our technological world was made possible by the powerful microchip, which severed in one fell swoop the tethers that bounded the application to the external means of control. The microchip is the soul of functional sentience.

The power chip is also the underwriter of the remaining tectonic sinews on our list. It is the brain behind artificial intelligence, the master key for blockchains, the connective tissue of the energy lattice, and the foot soldier of the internet of things. The invisible hand of the microchip gives form to algorithms, without which they'd revert to being mere playthings of computer scientists.

The great news story about microchips is its availability to business. The chip ecosystem is global in scale but local in presence, making the economics of custom design *and* low volume production compatible with any business model. Even small start-up companies can pursue their technology development initiatives with the knowledge that they can custom design the chips needed to run the innovation as it is meant to run, rather than compromise or sacrifice functionalities to make do with what is affordable (which was the general case until the early 2000s, surprisingly enough). In other words, the miracle prowess of the microchip is now available to the world, rather than the world of a few.

## The Niche Plays

Five niche plays are particularly noteworthy from a transformation standpoint: robots, *edge computing*, field simulations, local AI, and quantum computing, which is the "nichest" of them all. The first niche, robots, has been around for a very long time. It is only recently that they have merged into the killer app category of "intelligent automatas". Industrial robots have broken their static production floor shackles and literally walked out into the open. They have taken to the sea, the streets, and to the air as drones. They go about their autonomous business equipped with pattern recognition capabilities (see AI below), real-time interactions with control centers, and exquisite control systems that enable them to ambulate on any terrain with creepy, ersatz human mannerism (as Boston Dynamics have amply demonstrated—see www. bostondynamics.com). Robots incarnated themselves into self-driving cars, warehouse materiel handling specialists, micro surgeons, and nano-travelers of the human body. The next step is just on the doorstep, whereby these machines will be integrated in machine networks connected by the internet of things. The resulting level of autonomous *decision-making*, self-monitoring, and seamless collaboration with other devices will profoundly impact how companies organize their daily operations, and their interactions with regulatory bodies. The shape of these changes is hard to tell right now, since everything is evolving in parallel and in multiple directions. The state of flux can be unnerving; on the other hand, it augurs a world of innovation opportunities for the explorers of the AI realm. The opportunities in this case are not limited to technology either. The rise of autonomous machines brings about big picture questions in the ethical, societal, legal, insurance, and taxation spheres. Here is one instance where an inherently technological source of innovation gives rise to multiple innovation opportunities in the field of the humanities. By an ironic twist of faith, these opportunities will in turn reinforce the symbiotic relationship between data, machines, and humans—another sign of the arrival of the *Symbiocene* age.

The second niche is *edge computing*. Power chips have made functional sentience mobile through robotics. They have also shipped computing prowess from bulky servers to minute field devices. *Edge computing* refers to the design approach of installing data-handling algorithms at the source, whether on sensors or devices or instruments. In the past, these small and miniature systems collected and transmitted data only, leaving number crunching and decision-making to be performed by distant centralized servers and data processors. Today, the number crunching is embedded onto the device through which data transit. The ability to perform high-end calculations *in situ* became possible once power chips were small enough (with correspondingly low power consumptions). Edge computing is making lowly devices locally powerful. When assembled together into something like a dedicated internet-of-thing network, edge computing morphs into a functional sentient system. Edge computing is only just beginning too: the computing expectations of next generation autonomous systems (like a self-driving car) will not be delivered by one almighty onboard computer powered by advanced artificial intelligence; instead, the artificial intelligence operations will be divvied up between myriad sensors and devices equipped with stalwart edge computing features. *Edge computing will be to high-end machine autonomy what delegation is to large-scale corporate management.*

Field simulations are the third niche. Computer simulations are nothing new: powerful commercial software solutions have been around for decades to model the physics and chemistry of complex phenomena.* The underlying mathematical foundation of these software is relatively straightforward, and amenable to being run on a low-end laptop *when the object is simple and the physics linear.* Problems arise quickly with geometric or physics

---

* Often, these high-end software solutions will utilize the mathematics of finite element models to solve problems governed by coupled partial differential equations. These equations in turn formulate the transformation rules of the fields generated by the physical phenomena at play. For example, the flapping of a flag (a solid object with non-linear elastic behavior) from high winds (turbulent transient fluid mechanic) is a problem that can only be solved by finite element methods.

complexity or both. The computing hardware requirements grow exponentially with complexity. Hence, while the math is the same, solving it is the deal breaker. Simply put, reliable field simulations demand lots of computer memory, computing power, and power consumption. Fortunately, power chips with exotic computing capabilities have found their way into motherboards, memory boards, and video cards all at the same time. The result is extreme solver capabilities inside very small footprints (consider a $10 000 PC in 2018 can solve a problem that necessitated a $100 to $150 000 machine in 2008). In fact, this kind of solver capability is opening up the world of *affordable* simulations to manufacturers and design offices far and wide, across all industries, without breaking the bank. Simulations, in turn, vastly reduce engineering and design cycles; cut development costs by orders of magnitude; enable rapid investigations of "what-if" scenarios; increase reliability of the final products; and enable operators to determine quantitatively the design and the safe and unsafe operating envelopes of a system or a system of systems. Finally, when a problem is still too large to solve on a local machine, the availability of power chips organized into solver farms makes it possible for small companies to "rent" the solution power and compete on an equal footing with the largest conglomerates (who were, traditionally, the only firms able to afford to solve these problems).

> Field simulations running on power chips have democratized the economics of physics. Figuring out the inner workings of anything is available to anyone, anywhere, anytime.

The fourth niche application is local AI, which is a corollary of *edge computing*. In the latter, computing is tacitly taken to be the classical kind, i.e., an algorithm setting out a precise set of code instructions to be performed in a preordained sequence invariably and repetitively. AI on the other hand, operates without a prescribed codification, when the problem to be solved boils down to pattern recognition and/or inference. At the moment, AI applications are not transposable into a localized solution performed by a sensor or device. However, the physical algorithmic limitations will eventually be overcome, at which time sensors will be able to recognize patterns by themselves. An early indicator of this trend is the nascent arrival of portable language translation devices on the market. Once local AI is enabled, a network of sensors and devices will be inherently capable of broad functional sentience.

The final niche application is quantum computing. Quantum computing is to AI what fusion is to nuclear energy: the moonshot that always seems to be ten years away from reality. The field is a rarefied one and not one for the neophyte. The concept itself is complicated and exploits the weirdness of quantum mechanics (namely superposition and entanglement) to solve problems that are practically out of reach to classic binary computers (based on transistors that operate as on-off switches (1 or 0) to encode electrical signals into binary numbers— our common bits and bytes). Quantum computers do no such thing. The encoding of information is done with qbits representing superimposed quantum states. The qbit contains all probable solutions to a problem simultaneously. Specialized algorithms exploit this phenomenon, and some have already solved a class of problems, called NP, faster than any possible probabilistic classical algorithm run on a binary machine.* The field has been around for decades but has only recently been privy to successful demonstration experiments. Primitive commercial solutions have started to appear on the market. They are, for now, too immature to be offered to the general public. Once they are, however, huge classes of arduous problems will be accessible for the very first time, with cryptography and the science of passwords primed for a revolution.

# Artificial Intelligence

## *The Game Changer*

When Google decided to reincarnate itself into Alphabet, it did so with artificial intelligence (AI) as its anchor. This was a milestone in the early history of the Symbiocene age. It heralded the realization that the world was changing permanently. As Google's chief executive, Sundar Pichai, stated on numerous occasions, AI was the future of Google, one in which the company's offerings "would no longer represent the fruits of traditional computer programming, exactly, but machine learning".† Of course, Google is not alone in this school of thought. Virtually every major corporation on the planet has jumped on this bandwagon, from Facebook to GE and Apple to

---

* Quantum computing is no easier to understand than quantum mechanics. There are, however, genuinely pragmatic problems that can be solved only by quantum computing. See, for example: Hartnett, Kevin. Finally, a problem that only quantum computers will ever be able to solve. *Quanta Magazine*. 2018. Available at: www.quantamagazine.org/finally-a-problem-that-only-quantum-computers-will-ever-be-able-to-solve-20180621/.

† See also Gideon Lewis-Kraus' insightful article *The Great A.I. Awakening*, published by the *New York Times* magazine, 14th December 2016. Available at: https://www.nytimes.com/2016/12/14/magazine/the-great-ai-awakening.html.

Morgan Stanley. AIs are everywhere in the news, on the roads, in the living room. Self-driving cars, voice recognition, image identification, computerized medical diagnostics, car maintenance, network routings, sale projections, and future trends have been forced to pay homage to their computing overlord. AI has indeed become so prevalent that its acronym is becoming the name. However, like so many instances of technological wonderment, the expression artificial intelligence is easier said than understood. It can be as complex as rocket science.

The science of "artificial intelligence" began in the 1950s and has evolved by fits and starts along dramatically different theoretical directions. But what exactly is AI? It is quintessentially a software application powered by astounding hardware. It is not, however, a numeric form of intelligence in a biological sense. When asking about degrees of intelligence between an ant, a mouse, a dog, a raven, an elephant, a dolphin, and an orangutan, we tacitly presuppose some ability to solve various problems in one's environment with a second ability to adapt these solutions to new problems. Language and numeracy are another potent way of addressing the question. When these considerations are mapped to the computing machine, the answers would appear, at first sight, to be similar. However, this is mere mimicry. AI machines do not think; they deal with patterns. They rely on specialized algorithms to associate specific patterns to specific inputs (like detecting eyes in pictures of faces); to recognize these "learned" patterns when presented with new inputs; and to infer new correlations between sets of patterns (either learned or recognized). AI machines are programmed to deal with patterns, whereas non-AI machines are programmed to execute specific lines of codes according to pre-defined conditional sequences. That is why, for example, Google Maps is not AI but IBM's Watson machine, is. Much has been written about the dethroning of the human mind in the games of chess and go. However, these contests were profoundly flawed from a comparative standpoint. In order to defeat chess' world champion Gary Kasparov way back in 1997 (yes, that far back), IBM's Deep Blue machine had to be trained by its keepers on millions of game simulations, from which it was able to "learn" what series of moves are best for any specific board configuration. The proper and more telling test would have been to limit the training of Deep Blue to the same number of games that Kasparov had played in his life. Or better yet, introduce in the middle of a game an arbitrary change to the permissible move of one or more pieces. Without additional learning simulations, it is highly unlikely that Watson would have defeated Kasparov, who would have been to adapt on the fly, in real time, without flinching.

Chess notwithstanding, one cannot help but admire the abilities of AI with face recognition, fault detection, emotional reading, lie detection, language translation, natural language interactions with humans, medical diagnoses, legal reviews, data mining, buying trends, traffic debottlenecking, autonomous vehicles; the list of wondrous applications is seemingly endless. AI is also a category of software that encompasses many specialized applications such as neural networks, deep learning, data science, data lakes, back propagation, convolutional neural networks, machine learning, edge computing, and reinforcement learning. Underlying this plethora of specialties is a programming methodology that replicates the workings of the human brain, albeit on a vastly smaller scale. Hence the expression *neural network*. The brain comprises 100 billion or so neurons, with each neuron connected to perhaps 10 000 other neurons. That adds up to something between $10^{14}$ and $10^{15}$ possible connections! This is too large a number to replicate by the current state of the art in power chips. But the neural principle still works at much smaller machine scales. The connections are organized into layers, starting with the input layer. The inputs are processed successively by each layer. Each layer is programmed to focus on a narrow aspect of the pattern recognition problem. Within a layer, each neuron is assigned a group of *weights* and a so-called *activation function* that governs the generation of its output. The layer can be "trained" to recognize specific inputs by tweaking the neurons' *weights* until the correct solution is obtained—necessitating the mediations of a human. Networks constructed from several layers are called *deep networks*—the more layers, the more performing network.

The advantage of a neural structure stems from avoiding the need to create hard-wired connections within the network to store the outputs of each neuron (which would be impossible, in any case, when one is faced with $10^{12}$ or more connections). On the neural network, the sum of correlated outputs can be stored anywhere within its connections. The algorithm is trained by feeding it both the inputs (a picture for example) and the target output (the picture shows a dog). It is supplied with additional algorithms for tweaking, by itself, the weights of each neuron, until the final output matches the target. Then, the neural network is allowed to run its course until it achieves the correct match. The final set of neuron weights are then preserved in the memory for future references. This is what "training" is: a non-directed series of pathways within the network, traced by the algorithm and modified on the fly to correct the direction leading to the target.

## *To Think or Not to Think*

The essential consideration that matters to a digital management transformation lies at the systemic level. Neural networks are excruciatingly complex applications that require 1) massive datasets (the original killer app of "big data") to properly "train" the system; 2) sophisticated algorithms developed and operated by computer scientists (one does not improvise himself as a neural network operator); 3) correspondingly complex, large, and power hungry hardware to run the algorithms; and 4) a summary understanding, at the management level, of the abilities and limits of these systems to be able to ask the *right questions*. The dataset is by far the largest obstacle to running one's own AI department. The proficiency of a neural network is a direct function of the number of layers, the number of neurons in each layer, and the number of training input-output sets. The latter is the hardest to come by. A simple application like image recognition requires *billions* of images to properly train the network. How can anyone come up with such a vast array of inputs? The answer is not likely. Leading giants like Amazon, Apple, Baidu, Facebook, Google, IBM, and Google have an advantage over the rest of the world that cannot be matched. Their database holdings have the girth required to properly train their own networks. Joe Blow in his garage has no hope in hell of devising a packaged neural network system.

> Neural networks will produce astonishing results when properly trained and fed properly formulated questions. But their prowess is utterly useless if an analysis requires reasoning.

AI cannot think, or apply the scientific method by themselves, or perform long-term planning no matter how much data are thrown at them. The reason is mathematically simple: neural networks are, fundamentally, a linear sequence of continuous geometric transformations that map one vector space into another. AI systems are fundamentally subject to Gödel's incompleteness theorems. They cannot operate in a non-linear, out-of-sequence manner. Adding more layers, more datum inputs or more computing power will not overcome the non-linearity limitation. Given the computing heft required to train and operate an AI machine, one may ask whether getting into the game is worth the time and money. For one thing, its probabilistic nature makes it unsuitable when certainty of outcome is demanded. In the end, the decision to pursue an AI innovation project is best made from the constraints shown in Figure 2.1.

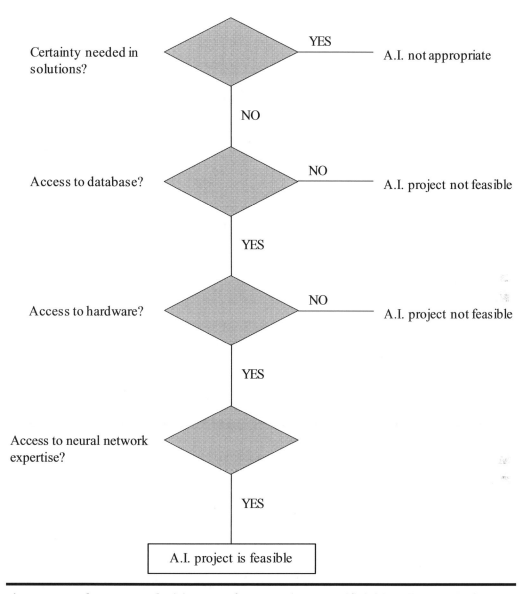

**Figure 2.1    The gonogo decision-tree for AI projects. Artificial intelligence projects are uniquely constrained in the innovation space. Faith, hope, funding, or visions are not enough to succeed; one must first reconcile the facts of his reality to the reality of the AI facts.**

## The Niche Plays

Five application subsets are associated with artificial intelligence: big data, materials, medicine, complexity, and job automation. As the name indicates, big data is concerned with very large sets of data. *From the business perspective*, big data is essentially a means to an end, which is to extract the patterns it hides and infer the intelligence insights to improve a firm's profitability. Big data is only useful if it can demonstrate a strategic advantage such as growing top line revenues, enable organizational transformations, enhance customer experience, uncover new customers and new markets, inspire new products and services, and assist management in making smarter decisions faster.* The promises of big data cannot be realized, however, by attempting to turn an existing information infrastructure into a big-data powered framework. Patches, workarounds, and band aid solutions slapped on a firm's legacy systems are a complete waste of money. But when done correctly (pursuant to the next two chapters), big data can deliver immediately to the bottom line in a way that few other management transformations can.

The materials niche is on its way to becoming a standalone field of computer science. The field also goes under the sexy name "material informatics". AI is enabling companies to abandon the costly trial-and-error approach to material science that has been the staple of all material discoveries for centuries. Costly, lengthy, labor intensive, sporadic, unpredictable, and serendipitous, the old ways could never be pursued on an industrial scale by anybody but large, government-funded organizations, yet still deliver results that were incomplete or unquantified. With AI, a definitive revolution is taking place the world over, led by little fish in an oceanic pond of entrenched bureaucracies. Outfits like Tide Materials Informatics, Exabyte.io, Citrine Informatics, VTT Proper Tune, Nutonian, Intermolecular, Alphastar, and Ques Tek Innovations, to name a few. These and other firms are the leading edge for the development of high-performance computing platform for materials discovery, often allying data mining, neural networks, and quantum mechanics. These sophisticated modules are integrated into potent software platforms capable of identifying hitherto unknown or unsuspected materials, and able to produce results within hours or days. Materials can be organic, inorganic, metallic, ceramics, photovoltaic, and polymers, to name but a few. Furthermore, these material platforms can also guide researches toward specific manufacturing processes to create these materials, from which production methods and equipment can be designed in record time.

The most famous production method is of course 3D printing, formally known as additive manufacturing. Complex geometries that are impossible with traditional methods (casting, forging, milling, machining, welding, etc.) can be created in a single process by placing exact amounts of materials at

---

* See *Driving Digital*, p. XV, by Isaac Sacolick (2017, American Management Association).

precise coordinates, curing them into solids, then repeating the process in successive layers until the entire shape is done. Early systems were limited to plastics and polymers to create small demonstration prototypes. Nowadays, metals, ceramics, rubbers, and concrete are available to industrial machines that can turn out sales-ready products; 3D printing is making it possible to manifest into the physical world what could only exist in an engineer's or an artist's mind.

The spectacular results obtained by additive manufacturing belie a more profound impact to the big manufacturing picture. Additive manufacturing has already led to 50% reductions in energy consumptions of manufacturing processes, and 90% reductions in source material costs. Just on this twin basis, additive manufacturing is proving to be a game changer across industries and design offices. It is fast becoming a cornerstone of the innovation process as well, given its ability to spin out working prototypes in minutes and hours, instead of days or weeks for traditional methods. Nevertheless, the real seminal transformation promise is ahead of us: the ability to manufacture *in situ*, within an operational setting, whatever replacement parts are needed ahead of an anticipated problem. Getting there, which is a matter of years, will require innovations in the areas of compensatory 3D data capture (whereby a worn part is scanned in 3D and adjusted automatically to compensate for deformations and material losses); better surface finish tolerances; simultaneous insertion of multiple materials; and autonomous final inspections and corrections. These and other functionalities will be driven by new AI algorithms coupled to new hardware designs. Once there, additive manufacturing will become a mission-critical component of the digital-industrial paradigm, discussed later in this chapter.

From the business perspective, the materials niche represents a great potential to solve intractable design problems such as corrosion and erosion resistance, pollution control, high efficiency energy converters, renewable energy, lightweight engineering materials, and electric charge storage (among myriad others). A firm needs not become a material informatics operator. The beauty of the thing is that we are entering an era of made-to-order custom materials, a nascent consulting domain akin to chip foundries, outsourced manufacturing, and electronic contract fabrication.

> AI is giving rise to a new business model where material properties can be ordered on spec to fit a specific design requirement of a product. Material properties are now becoming commercial assets in their own rights.

The fourth niche, medicine, is rapidly transforming the practice of medical practice everywhere. Specialists and researchers can use AI to review enormous databases of research papers and studies, then infer from them statistically significant correlations between causes and effects of disparate treatments. AI is also superbly efficient at rendering diagnostics from x-rays and lab tests, in some instances at a higher rate of correctness than human specialists. AI can detect, again through pattern recognition, side effects and unsuspecting growths that warrant closer inspections. AI, in other words, is rapidly becoming a doctor's trusted assistant/second opinion expert. Precision surgery, remote-controlled microtools, and non-invasive surgery at the cell level are next on the horizon.

Complexity is the cousin of the field simulation niche. Whereas field simulations exploit the prowess of power chips to quantify the physics of a phenomenon under study, complexity simulations exploit the power of AI to recognize the transient patterns of dynamical systems like crowds, road traffic flows, foot traffic in airport terminals, power grid load distribution, and so on. These systems are inherently complex, in both mathematical and organic sense, but outside of closed-form mathematical rules of behavior. From the business perspective, the real-time control of these systems becomes paramount when they are mission critical—say, the coordination of thousands of workers on a large construction site, or a what-if scenario if an explosion occurs in a mall. Detecting the onset of a disturbance to the natural "flow" of these systems becomes a profit-loss issue with immediate impact to the bottom line. AI is ideally suited to take on the role of pattern monitor and advance warning system when things look to get unhinged.

The final niche is job automation. Ever since machines came on the scene, humans have lost the battle when efficiency, reliability, and repetition are in play. With AI invading the driver seats of truckers, taxi drivers, and train conductors, a new luddite threat is appearing on the employment horizon. AI is making advances even into the elitist realms of the medical, legal, engineering, and management professions. This trend is underway and will not stop, if only because it is driven forward by cost reductions, liability and warranty limitations, proficiency, productivity, and labor pressures. Shrinking demographics across the world are working against workers but for this trend, unfortunately (and counter-intuitively).*

---

* See the report by the consultancy Bain and Company, *Labor 2030: The Collision of Demographics, Automation and Inequality*. Available at: https://www.bain.com/insights/labor-2030-the-collision-of-demographics-automation-and-inequality.

From the digital management transformation viewpoint, this niche can become a pyrrhic contest. Automation, when properly done, will yield efficiency and productivity gains in the short term, with a competitive advantage over the competition as a bonus. Over longer timescales, however, automation leads to profit stagnation. Once it has spread across competitors, the productivity advantage vanishes. Widespread automation also *decreases* a firm's ability to pursue continuous improvements and new cost cutting opportunities, since machines are sunk costs with enormous operational inertia. One simply does not replace a fully automated production line with the latest and greatest technology advances whenever they show up. With automation, the process becomes an entrenched status quo, which leads to the fifth paradox:

> Paradox 5: The proliferation of automated processes and systems causes a generalized loss of flexibility and competitive advantage by the firm, five to ten years hence.

The corollary of this paradox affects a firm's talent pool. Each automate process eliminates the human connection to the basis of that process. The loss of that connection leads to a loss of understanding, by the firm, of the now hidden constraints, limitations, and inefficiencies of that process *relative to the bigger picture of the overall operation*. And here is the hic: at the moment, there is no AI application that could replace the human ability to gauge these hidden defaults of an automated process. In other words, the firm loses five precious intangible strategic assets when it eliminates the human connection to the automation of a process:

1. The knowledge asset into the inner workings of the process
2. The performance assessment of that process
3. The understanding of the ramifications of that process upon the whole system
4. The human insight into the opportunity to improve the whole system at the process level
5. The human expertise and process skillset

As if this was not enough, the automation juggernaut is afflicted by the sixth digital paradox:

> Paradox 6: Algorithm, know thyself. Automation through AI yields outcomes bereft of the rationalization behind them. AI offers the "what" without knowing the underlying "why".

The question that dominates AI research at the moment revolves around the invisibility of the inference process adopted by an AI system to produce an answer. The issue of "why" an outcome was produced among all others becomes critical when one wishes to learn from it, establish a precedent from it, or understand how the outcome missed the boat. The salient example comes to us courtesy of the legal profession. Say an AI application is used to analyze the pertinent precedents to a case in front of a judge. Whatever the conclusion is offered by the application, the judge, the lawyers, and the parties to the judgment will want to know the basis for the outcome, if only to decide whether or not an appeal is warranted. Unfortunately, at this time, there is no way to tell why AI produces a specific outcome when the inputs are characterized by amorphous features like opinions rather than quantifiable numbers.

# Blockchains

## *The Superstar Neologisms*

The year 2008 is forever associated with the global financial meltdown. It was also the year that gave the world two of its more recent and famous neologisms: bitcoin and blockchain, formally known as *distributed ledger technology* (DLT). The blockchain is poised to revolutionize the way people, organizations, and governments establish the legitimacy of transactions. Conceptually, a blockchain is a database whose records are distributed across a network, rather than centralized in one place, and visible to *all* participants on the network. All records and transaction instances between participants are visible to every other participant. The details of each transaction are only known by the parties privy to the transaction. The verification of the legitimacy of the transaction is performed simultaneously by participants on the network, making the system tamper and hack proof. Each *record* on the database is called a block, uniquely identified, validated, and never erased. Blocks are assembled sequentially as a chain, which form the "ledger". It is this ledger that is visible to everyone on the DLT network.

The blockchain is theoretically impossible to hack by intent or by collusion. If a change is made, that change is seen by *everyone else whose name appears in at least one block.* In accounting parlance, the blockchain is said to be an open, transparent, and shared ledger that documents the existence of transactions in a permanent, verifiable manner. The beauty of the DLT is its ability to function *without a centralized authority over the DLT* (cutting out costly middlemen), and create a transactional framework that is arrantly transparent, traceable, and impervious to bookkeeping shenanigans. Take for example the banking system, a bastion of centralizing proclivities. In the traditional setup, all financial transactions are vetted by a single authority (the clearing house). This clearing house exerts monopolistic power over the entire network. This singularity makes the system vulnerable, since it offers a single point of entry by malfeasance mercenaries. Within a DLT ecosystem, each participant is both actor and clearing house authority at the same time. There is no central authority to decide anything. The legitimacy of every transaction is independently verified by *all* participants concomitantly. Hacking the ledger becomes impossible in real time.

## The Hacker-Free Ledger

The blockchain is, again, a ledger (in a pure accounting sense) that is operated as a highly distributed computing platform (shown in Figure 2.2 and characterized by a "Byzantine fault tolerance" capability).* The blockchain gives rise to a decentralized consensus that is always up-to-date, acting in an evidentiary capacity regarding the legitimacy of its records. All records are seen by all at all times; none is hidden nor can it be hidden. Therein lies the fantastic potential of the system: the blockchain is the only system capable of maintaining record legitimacy in the digital space. The blockchain did for digital record management what double-entry accounting did to accounting in the 14th century: it revolutionized the practice.† The great corporate malfeasance scandals by Enron and Bernie Madoff could never have been possible had a blockchain been in use at the time.

The blockchain is much more than bitcoins. DLT has already become mission critical in industries such as logistics and transportation, mining

---

* Byzantine fault tolerance (BFT) is the ability of a computing system to remain dependable despite component failures or imperfect inputs. The expression "Byzantine failure" refers to a component that appears to be both failed and functioning, in an inconsistent manner, thus being dreadfully hard to troubleshoot. The expression has its roots in the so-called Byzantine Generals' Problem, where actors must agree on a concerted strategy to avoid catastrophic system failure while some actors are not reliable. See: Wikipedia. Byzantine fault. Available at: https://en.wikipedia.org/wiki/Byzantine_fault. Last modified 30th March 2020.

† The oldest record of a complete double-entry system is the accounts of the Treasurer of the Republic of Genoa, circa 1340.

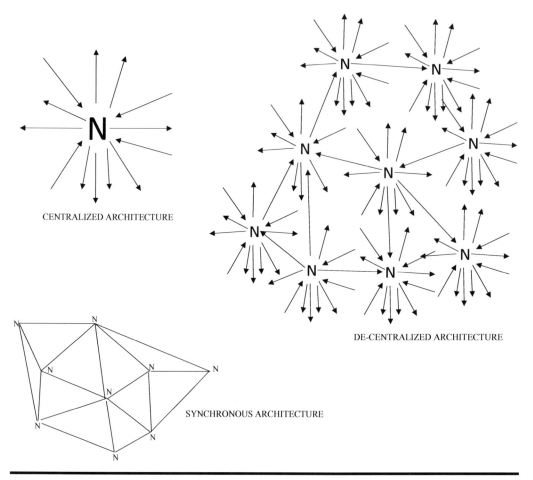

CENTRALIZED ARCHITECTURE

DE-CENTRALIZED ARCHITECTURE

SYNCHRONOUS ARCHITECTURE

**Figure 2.2   Distinguishing features of the DLT network. Historically, large databases have been centrally operated, as shown on the left. The internet broke this model and decentralized the network into a set of regional hubs (middle image). The blockchain does away completely with any node or hub (right image). The internet is still there but running as the plumbing beneath the DLT.**

(and mining rights), material traceability, peer-to-peer trading, government services, digital currencies, and individual health records, to name but a few. So-called smart contracts are also finding their way into the routine management of commercial contracts. The list of possible applications is endless. Which begs the question: when is a blockchain framework commercially viable? This is an important issue since DLT are unparalleled in specific applications but useless (and complex and costly) in others. Not every business process is suited, or suitable to a DLT operation. The latter will make sense when the following business imperatives MUST be met:

- Provable, instantaneous traceability of a transaction (for example, to ensure that a drug has been manufactured, tested, shipped, received, and administered correctly and legitimately)
- Members of a supply chain are legitimate, qualified, and required to maintain accurate documentation for auditing and regulatory compliance
- Every step of a contractual engagement must be visible, auditable, and certifiable
- The requirements above must be met by numerous participants

The last imperative is the primary driver. If a business deals with a small number of vendors and clients, the other three requirements could be satisfied without the implementation of a DLT framework. Practically, this means one or two people able to manage the participants by spreadsheet. Anything beyond that scale will explode the complexity of the management process, at which point a DLT framework will be justifiable. Then comes the question of DLT design and implementation. The field is still in its infancy and attracting thousands of eager innovators wanting a piece of the pie. One can compare the state of the industry today with the browser wars in the early days of the internet (remember Napster, AOL, and Myspace?). There are no simple recipes or decision checklists to decide among the thousands of solutions offered today. From a digital management transformation standpoint, the issue is not which DLT framework to adopt (Hyperledger Fabric? Ethereum? Ripple? Bitcoin? Iota? Or something entirely new?); the issue is about defining whether or not a DLT is commercially *essential* to the firm's core operations. Everything after that is mere design detail. The benefits of a commercially essential DLT will deliver immediate impacts to the bottom line, mind you. The firm will be able to cut out costly middlemen and transaction costs dramatically. Precisely managed transactional information (coded as blocks on the chain) will enable the firm to find out what processes are *valunomic* vs *wasteful* (a concept explored in the next chapter). Warranty costs, user experience, and product quality will improve. Data monetization will be enabled. Regulatory compliance costs will shoot through the floor. Bulky and inefficient quality assurance labor will be dramatically shrunk.

The flurry of development and investment activities surrounding DLT make it impossible to nail down a limited list of niche applications in this case. Literally, each new deployment of a DLT solution to a new business function constitutes a niche on its own.

# The Energy Lattice

## *Dumping the C*

The world has gotten on board the renewable energy train. It is no longer a question of if or when, but how fast and how far. Clean energy technologies (clean at least when producing electricity but not so much when mined and manufactured) are spreading across nations faster than wildfires. Batteries are changing the economics of solar and wind power generation and are enabling electric motors to drive down the highways. Traditional energy sources (oil and gas, coal, hydroelectric, and nuclear) will not fade out anytime soon (except for coal, perhaps), which means that the planet will continue to exist in a state of power generation flux for several decades to come. The utopia of 100% renewable power is just that, a utopia that has zero chance of becoming reality. But the possibility of 100% clean energy will fare better.* Within two to three generations, the world may have decarbonized its energy diet. The end of C (as in Carbon) in the energy equation is coming, in what we will call the Great Dislocation.

The clean energy impetus is irreversibly launched, which will have profound ramifications on power distribution on this planet. Batteries are unlikely to be the dominant form of storage and will remain a transient evolutionary step on the likelier dominance of stored hydrogen (in gaseous and liquid forms). The reason is wrapped inside two words: *energy density.* Take a look at Table 2.1, which lists in alphabetical order a selected number of commonly available energy sources by energy content per unit of mass, and Table 2.2, which compares them by energy density.†

The conclusion is inescapable: battery-based systems cannot, today or in the future, replace hydrocarbons. The future lies with hydroelectric, hydrogen, and nuclear sources (fission and fusion). Apostles of the renewable creed (who fanatically embrace solar, wind, ocean waves, and geothermal sources against the limits of physics) cannot win the day. Physics and spatial limits will prevail with utter disregard with renewables' articles of faith. Not convinced yet? Consider these inconvenient facts. The world's energy production profile bears this out factually. The top ten electric power plants in the world are either hydroelectric or nuclear, measured by capacity factor

---

* In 2016, the world consumed 560 exajoules (560 × $10^{18}$ joules) of energy. To produce this amount of energy from solar panels alone would require a land area equivalent to 6680 earth planets.

† Further reading on the topic of energy density is easily accessible at: Wikipedia. Energy density. Available at: https://en.wikipedia.org/wiki/Energy_density. Last modified 9th March 2020.

**Table 2.1   Energy Densities of Common Energy Sources**

| Storage Type | Mechanism of Energy Release | Energy Density | | Dower density | |
|---|---|---|---|---|---|
| | | MJ/kg | BTU/lb | kWhr/Kg | kWhr/lb |
| Battery, lithium-air rechargeable | Electrochemical | 9.0 | 3882 | 2.50 | 1.1 |
| Battery, zinc-air | Electrochemical | 1.6 | 686 | 0.44 | 0.2 |
| Biodiesel oil (vegetable oil) | Electrochemical | 42.2 | 18204 | 11.72 | 5.3 |
| Compressed Natural Gas (CNG) at 250 bars | Mechanical | 53.6 | 23121 | 14.89 | 6.8 |
| Coal, anthracite | Chemical | 60.0 | 25882 | 16.67 | 7.6 |
| Compressed air at 300 bar | Mechanical | 0.5 | 216 | 0.14 | 0.1 |
| Diesel | Chemical | 45.6 | 19670 | 12.67 | 5.8 |
| Gasohol E10 (10% ethanol 90% gasoline by volume) | Chemical | 43.5 | 18782 | 12.09 | 5.5 |
| Gasohol E85 (85% ethanol 15% gasoline by volume) | Chemical | 33.1 | 14278 | 9.19 | 4.2 |
| Gasoline | Chemical | 46.4 | 20015 | 12.89 | 5.9 |
| Hydrogen, at 690 bar and 15°C | Electrochemical or chemical | 141.9 | 61193 | 39.41 | 17.9 |
| Jet A aviation fuel/kerosene | Chemical | 42.8 | 18462 | 11.89 | 5.4 |
| Lithium borohydride | Electrochemical | 65.2 | 28125 | 18.11 | 8.2 |
| Liquified Natural Gas ( at −160 °C) | Chemical | 53.6 | 23121 | 14.89 | 6.8 |
| Methanol | Chemical | 19.7 | 8498 | 5.47 | 2.5 |
| Natural gas | Chemical | 53.6 | 23121 | 14.89 | 6.8 |

(Continued)

**Table 2.1 (Continued)   Energy Densities of Common Energy Sources**

| | | Energy Density | | Dower density | |
|---|---|---|---|---|---|
| Storage Type | Mechanism of Energy Release | MJ/kg | BTU/lb | kWhr/Kg | kWhr/ lb |
| Water at 100 m dam height\ | Mechanical | 0.001 | 0.43 | 0.00 | 0.0 |
| Wood | Chemical | 18.0 | 7765 | 5.00 | 2.3 |
| **The limits of physics** | | | | | |
| Antimatter | Annihilation at contact with matter | 9.0E+10 | 3.9E+13 | 2.5E+10 | 1.1E+10 |
| Deuterium | Nuclear fusion | 8.8E+07 | 3.8E+10 | 2.4E+07 | 1.1E+07 |
| Plutonium-239 | Nuclear fission | 3.1E+07 | 1.3E+10 | 8.6E+06 | 3.9E+06 |
| Uranium | Nuclear fission | 8.1E+07 | 3.5E+10 | 2.2E+07 | 1.0E+07 |
| Thorium | Nuclear fission | 7.9E+07 | 3.4E+10 | 2.2E+07 | 1.0E+07 |
| Plutonium-238 | Radioisotope thermoelectric generator - used in deep space probes and satellites | 2.2E+06 | 9.7E+08 | 6.2E+05 | 2.8E+05 |

*Note:* This table compares the energy contents of energy sources by mass. The densities exclude all other considerations.

(the total energy produced in a year, which amounts to 8766 hours). No facility can operate continuously over 8766 hours. Maintenance, shutdowns, accidents, labor strikes, source shortages (nighttime for solar panels for instance) mean that a plant will reduce the capacity factor from 100% (8766 hours of production) to a much lower number. According to the US Energy Information Administration,* typical capacity factors for North American facilities in 2016 were 92% for nuclear plants, 55% for combined-cycle natural gas plants, 38% for hydroelectric plants, 35% for utility-scale wind farms, and 27% for utility-scale solar photovoltaic installations. The Itaipu dam in Brazil took first spot in 2016 for both capacity factor (84%) and maximum energy produced (103 billion kilowatt-hours [kWh] or 0.3 exajoules). By way of comparison, the world's biggest solar array (at the time of writing of this book)

---

* Conca, James. The biggest power plants in the world: Hydro and nuclear. *Forbes.* 2017. Available at: https://www.forbes.com/sites/jamesconca/2017/08/10/the-biggest-power-plants-in-the-world-hydro -and-nuclear/#2c84b95f2c88.

**Table 2.2 The Futility of C-Free Sources**

| Storage Type | Mechanism of Energy Release | Energy Density | | Power Density | |
|---|---|---|---|---|---|
| | | MJ/kg | BTU/lb | kWhr/Kg | kWhr/lb |
| Water at 100 m dam height | Mechanical | 0.001 | 0.43 | 0.00 | 0.0 |
| Compressed air at 300 bar | Mechanical | 0.5 | 216 | 0.14 | 0.1 |
| Battery, zinc-air | Electrochemical | 1.6 | 686 | 0.44 | 0.2 |
| Battery, lithium-air rechargeable | Electrochemical | 9.0 | 3882 | 2.50 | 1.1 |
| Wood | Chemical | 18.0 | 7765 | 5.00 | 2.3 |
| Methanol | Chemical | 19.7 | 8498 | 5.47 | 2.5 |
| Gasohol E85 (85% ethanol 15% gasoline by volume) | Chemical | 33.1 | 14278 | 9.19 | 4.2 |
| Biodiesel oil (vegetable oil) | Electrochemical | 42.2 | 18204 | 11.72 | 5.3 |
| Jet A aviation fuel/kerosene | Chemical | 42.8 | 18462 | 11.89 | 5.4 |
| Gasohol E10 (10% ethanol 90% gasoline by volume) | Chemical | 43.5 | 18782 | 12.09 | 5.5 |
| Diesel | Chemical | 45.6 | 19670 | 12.67 | 5.8 |
| Gasoline (petrol) | Chemical | 46.4 | 20015 | 12.89 | 5.9 |
| Compressed Natural Gas (CNG) at 250 bars | Mechanical | 53.6 | 23121 | 14.89 | 6.8 |
| LNG (NG at −160 °C) | Chemical | 53.6 | 23121 | 14.89 | 6.8 |
| Natural gas | Chemical | 53.6 | 23121 | 14.89 | 6.8 |
| Coal, anthracite | Chemical | 60.0 | 25882 | 16.67 | 7.6 |

(Continued)

**Table 2.2 (Continued)    The Futility of C-Free Sources**

| | | Energy Density | | Power Density | |
|---|---|---|---|---|---|
| *Storage Type* | *Mechanism of Energy Release* | *MJ/kg* | *BTU/lb* | *kWhr/Kg* | *kWhr/lb* |
| Lithium borohydride | Electrochemical | 65.2 | 28125 | 18.11 | 8.2 |
| Hydrogen, at 690 bar and 15°C | Electrochemical or chemical | 141.9 | 61193 | 39.41 | 17.9 |
| The limits of physics | | | | | |
| Plutonium-238 | Radioisotope thermoelectric generator—used in deep space probes and satellites | 2.2E+06 | 9.7E+08 | 6.2E+05 | 2.8E+05 |
| Plutonium-239 | Nuclear fission | 3.1E+07 | 1.3E+10 | 8.6E+06 | 3.9E+06 |
| Thorium | Nuclear fission | 7.9E+07 | 3.4E+10 | 2.2E+07 | 1.0E+07 |
| Uranium | Nuclear fission | 8.1E+07 | 3.5E+10 | 2.2E+07 | 1.0E+07 |
| Deuterium | Nuclear fusion | 8.8E+07 | 3.8E+10 | 2.4E+07 | 1.1E+07 |
| Antimatter | Annihilation at contact with matter | 9.0E+10 | 3.9E+13 | 2.5E+10 | 1.1E+10 |

*Note:* The woeful performance of battery-based renewable systems (mainly solar and aeolian) is blatantly obvious.

is the Kurnool Ultra Mega Solar Park in India. It produced 2 billion kWh in 2016, over a spatial footprint of 24 km² [9 miles²]). *This is a hundred times less than the top producing* plant. The largest wind farm is in Gansu, China. It produced 24 billion kWh of electricity in 2016, over a footprint of 50 km² (19 miles²), enough to power Los Angeles. Finally, remark that the top ten power plants produced a total of 0.53 exajoules, which amounts to 1% of the world's energy consumption in 2016 (560 exajoules). Those two numbers explain why the utopian dream of going 100% renewable in energy production is just that, utopia. It is physically impossible to achieve them. Moreover, let us not forget the other consequence of a renewable paradigm: the capital investment in mining and transportation infrastructures that must be diverted from resource exploration (hydrocarbons and water) toward *mining* operations to produce the raw materials of renewable technologies—indium, tellurium, and gallium for solar panels; lithium, cobalt, and manganese for batteries; and rare earths for high-performance magnets in motors—along with the environmental ramifications that they entail.

Which means that the greater business opportunity lies not with renewable energies but in extracting further efficiencies from traditional sources.

## To the Infinite Grid and Beyond!

What interest us at this point is not so much the particular solutions that will come out of the Great Dislocation but their farther-reaching impact on the energy distribution networks. We can draw an interesting parallel between the latter and the blockchain. Power distribution has hitherto been a centralized affair, with few but large power plants producing all electricity for large regions. Some European countries, Germany notably, have forced their power grids to open up to locally produced electricity from small-scale solar panels. Their grids have slowly begun to transform themselves into decentralized distribution networks. This trend is irreversible worldwide: if renewable energies are to flourish, centralized networks must evolve towards de-centralization. And this is indeed what can be witnessed on all continents. Power grids are loosening their grip on power (pun intended). This transition to a decentralized modus operandi is made possible by... power chips, AI, and DLT (surprise surprise...).

The end game however is further out. This will be when those grids are fully distributed, with power flowing back and forth between local producers/consumers. There is more to this future configuration than simply plugging into a grid and drawing energy from it. The localization of the energy nexus (supply AND demand) will dramatically alter the technological landscape as well. For all the variables (inputs and outputs) that factor in the operation of a large power plant also appear at the smallest of scale. Things like emissions, heat rejection and recovery, noise, local storage, *in situ* datum monitoring, internal spot performance tracking, and autonomous machine decision-making will become routine, nay necessary components of localized power installations. Large industrial facilities will in fact become self-administered power networks in their own right, superimposed in real-time onto the regional grid beyond the fences.

These myriad concepts are the reason why we speak of the energy lattice rather than the grid. The grid limits the perspective to the hardware involved with power distribution. The lattice injects a level of conceptual abstraction into the picture by elevating the inner workings of the grid to that of a living organism, with all the extra dimensions that it entails.

## Business Opportunities Galore

The digital management transformation opportunities that lie in wait augur excitedly on the foreseeable horizon. At the technology level, everything remains to be invented. Common installation solutions provide a starting point but have yet to exhibit the capabilities to operate in a fully distributed framework. Even transient batteries will be part of the scene for the foreseeable future. Hydrogen storage and handling is essentially virgin territory in this application space. Power from heat, produced or from waste, has barely scratched the innovation surface. At the control level, the field is equally wide open. Autonomous power production and management will necessitate step changes in control systems and machine-based governance digital structures. *In situ* decision-making will likely involve some form of artificial intelligence system, which will need to connect to the outside world concomitantly. Finally, there is the integration opportunity, whereby all these bits and pieces and algorithms must be installed, deployed, and monitored. All of this work is new, since none of the bits and pieces and algorithms are ready. The integration is, in fact, the greatest of them all, as it ties into the next digital tectonic plate: the digital-industrial transformation.

# The Internet of Things (IoT)

## The Quiet Revolution

The internet of things is here. The difference with the plain old internet is users. Machines, devices, sensors, and algorithms will be the principal users of the former, while people and organizations will continue to be the prime users of the internet. Obviously, the internet backbone will continue in its role of plumbing to the flows of information on it and the IoT.

> The IoT can be summed up in one sentence: machines dealing with machines independently of humans.

The machine could be the simplest on-off switch or a monitoring device. It could be a sensor or a chemical treatment. It could be a Bluetooth transmitter or an entire control network. Whatever its nature, its size or its function, the machine is connected to the rest of the world's industrial network, exchanging information continuously in fantastic volumes. The transactions occur automatically and autonomously, without any interventions by humans. Five billion devices are estimated to be connected in this manner right now, a number that is expected to reach 30 billion within a decade, with a US$10 trillion windfall in profits and cost savings for their owners.*

## The Autonomous Paradigm

The IoT is much more than countless widgets connected together over the internet. It is a transformative paradigm shift where the machine network exists as a complex living organism, out of which will emerge *functional sentience*. The analogy is closer to reality than lyricism suggests. Take the human body for instance. It is a fantastically complex chemical factory. It is autonomous, self-correcting, self-healing, self-aware, and environment-analyzing. It is chemically powered, centrally controlled, and decentralized in actions. It is incredibly connected through a network comprising trillions of synaptic connections. Furthermore, it is self-replicating and able to manufacture its replacement parts on its own. The human body does all of this, and more, with only five essential inputs: oxygen, water, nutrients, DNA, and sleep. Five inputs that have given us you, the reader, and Curie and Raphael and Mozart.

Consider now a run-of-mill traditional factory for tires. From raw materials to finished products, 20 essential processes are involved. Each process is itself constituted of several steps, procedures, and quality control. Each such mechanic is in turn subject to a variety of datum collection, analyses, tracking and costing, and filing. If this factory is analog, bereft of sophisticated digital frameworks, the cost to operate is extensive human interventions. Each mechanic may require lots of humans making lots of decision continually to make sure that the steps are in sync, in the right quantities, and to the correct specifications. Needless to say, profitability and *valunomy* will suffer.

What might an IoT framework look like? Something like the right side of Figure 2.2, with each node representing one of the 20 primary processes. Each node is a network unto itself that is embedded into the bigger manufacturing network for the tire operation. Each *nodal network* runs independently of the others but remains in constant contact with the other nodal

* Source: Frost & Sullivan White Paper 2016. *Enhancing Business Productivity with the Internet of Things (IoT)*. Available at: https://www.rogers.com/cms/rogers-enterprise/page-specific/products-and-solutions/internet-of-things/resource-centre/business-productivity/pdfs/business-productivity-en.pdf.

networks through the information transaction that transits across the overall web. The connectivity between the nodal networks makes the entire operation digital. But the autonomous nature of each node, interacting in real time with all other nodes, is what makes the whole thing digital industrial.

## The Need for Speed

Each node in Figure 2.2 represents a localized instance of one or more threads discussed in this chapter, with a propensity toward *information dominance*. This information dominance underscores the monetization potential of data and the strategic ability for *valunomic decision-making*. Companies have hitherto been held captive by a reactive process in matters of operational effectiveness, owing entirely to the lack of actionable feedback from their daily operations. There is simply too much information produced at any given time for humans to wrap their brains around. So, companies have had to triage these mountains of information to whittle them down to manageable bits captured in some ersatz real-time fashion. Thus were born production reports, material inventories, productivity data, and time sheets. However, real-time analysis of those record sets remained for the most part impossible.

> The point of the IoT, at the management level, is to maximize real-time optimization of the firm's operations.

Optimization here takes on a specific meaning: *anticipatory management* of potential issues that threaten profitability. "Anticipatory" in this context equates "pro-active", as in initiating the decision-making process *before* an event makes it necessary. Management by anticipation requires live data, analyzed in real time, from which trends and patterns can be detected *before affecting materially any facet of the operations of the business*—bringing into the picture AI as an enabling mechanics of decision-making *by* humans. Managing by anticipation is far more demanding than by reaction (the usual *modus operandi*). But it yields immediate, tangible benefits in terms of failure containment, cost overruns, profit opportunities, and risk oversight.

## Delegation of Authority

Operationally, anticipatory management is only possible through a synchronized control schema. The efficacy of the structure lies in the built-in delegation of authority to the machine/sensor/device to act proactively at the level of the node. The trick, of course, is to design algorithms that can manifest this delegation of authority AND configure the physical aspects of the nodes to be able to implement the autonomous decision. This is no easy matter; by the same token, it represents a phenomenal opportunity for operational efficacy. Still greater benefits will accrue to the digital-industrial organization. The authority delegation builds the shortest bridge between the decision-making process and the conditions most impacted by the decision. The machine intervention can be instantaneous and autonomous (i.e., without human inputs), including the creation of replacement parts via *additive manufacturing* systems, for example. The net effects to the bottom line are significant:

- Asset reliability is increased across all nodal scales.
- Production throughput is maintained or optimized to the conditions on the ground.
- The firm's *digital network* provides a pervasive wireless connectivity to everything, everyone, and everywhere at all times. Anticipatory management by machines or by humans becomes the de facto *modus operandi.*
- The connectivity permits remote monitoring in real time at all nodal scales, which in turn improves security and compliance (regulatory, environmental, operational). Monitoring is performed through node *presence data*, smart video and sound, drone surveillance, and satellite tracking. AI applications are applied to perform image recognition, diagnostic predictions, and situation analyses in real time, autonomously.
- The connectivity supplies the power and the engine to enact constant predictions on equipment failures, leakages, downtime, bottlenecks, and other degraded conditions.
- Constant predictions feed the data required by asset integrity systems that manage, autonomously, maintenance programs in an anticipatory fashion, through real-time analytics.
- Mobile nodes (humans, vehicles, material handling, etc.) are equally monitored in real time, throughout the wireless network, for fleet efficiency, deployment optimization, problem detection, and geospatial tagging.
- Personnel safety, availability, and productivity can be measured through wireless real-time tracking, video analytics, and autonomous incident assessments.

- Global metrics can be compiled from aggregated nodal data and calculated intelligence information. Hidden trends, silent underlying currents, and optimization correlations can be inferred through plenipotentiary AI inference systems.

The combined effects of these benefits yield an order of magnitude improvement in overall operational effectiveness, cost valunomy, personnel optimization, and regulatory compliance. In other words, improved commercial performance and capital investment returns.

## *Transformation Strategy*

The IoT enables an organization to deconstruct its operation into granular processes and assess what works, what is senescent, what needs upgrading, and what must go. However, the end game of the transformation cannot be achieved merely through the addition of wireless routers, signal processors, fancy software, and costly server farms.

> An IoT strategy is not appended to an existing business framework; it requires a complete recast of it by digitizing it from A to Z.

The undertaking is neither easy nor cheap. It will be hard, the more so when the culture of the organization clings to the comforting fallacies of its past successes. Remember that we are entering a world where your customers, your supply chain channels, your competitors, your regulators, your products, your people, and your operations are going digital. These are seminal disruptions to the stodgy *status quo*. The magnitude of the information connectivity could be staggering. For the transformation strategy to succeed, its implementation cannot be limited to bolting on new bits and pieces and processes. It requires a tactical perspective that completely rethinks how the business will interact with the other species that populate this new ecosystem. Which may very well change what you sell and the value proposition that you offer.

Happily, enterprise-wide commercial solutions are coming to the fore by corporate giants (such as General Electric and Siemens for example). The complexity of the foundational framework is best left to the global industrial hegemons of this world. The aspiring digital transformation candidate should focus instead what it can and must control within its operational ecosystem.

# The Holobiome

## *The Digital Commercial Ecosystem*

The expressions "business ecosystems", "market ecosystems", and "supply chains" are commonly encountered in the literature to describe the particular environments in which a firm or organization operates. The ecosystem conveys the notion that the firm will be interacting with a multitude of dependent and independent entities such as competitors, labor spools, regulators, stakeholders, shareholders, and objectors/opponents. The continued commercial viability of the firm rests in large parts on its ability to manage these relationships. There can be several ecosystems operating in parallel, sometimes coterminously and sometimes in subduction. The terminology is not adequate to capture the convergence effects of the Symbiocene age, with the digital integration of machines, people, organizations, and governance bodies into a single human Pangaea plate.

This is the reason for introducing the neologism "holobiome". The word is a contraction of the words *holonomic* and *biome*. The former being a physical system whose state is a function of the path/evolution taken in order to achieve it. The latter describes a community whose inhabitants share characteristics for the environment in which they exist. The holobiome is thus an ecosystem comprising humans, machines, and their mutual interactions. The *holobiome* of a firm comprises the ensemble of common relationships defined by entities external/independent of the firm (vendors, buyers, regulators, etc.)—which are called *exogenic agents*; and a second ensemble of proprietary relationships arising from the internal workings of the firm (employees, business processes, management team, shareholders, etc.)—which are called *endogenic agents*. By extension, the agents are progenitors of *risks*—exogenic agents drive the risks that are not under control of the firm, while endogenous agents foster the risks that *are* (or should be) under the control of the firm.*

## *Understanding the Holobiome*

A successful digital management transformation cannot be brought to fruition without understanding a firm's relationship to its holobiome, as a consequence of the first digital paradox. Of course, the starting point of these relationships

---

* The delineation of risks into external and internal risks in this manner was originally suggested by the author in *Investment-Centric Project Management*.

is the datum making up the bloodstream of the information flows within the holobiome. Data become commands or controls or triggers or messengers of intent that are translated into physical actions by machines or humans or both. This complex choreography plays out in infinite ways by innumerable systems interconnected between separate holobiomes. Nevertheless, these information streams unto themselves are of no use without the decision-making required to direct them, giving rise to the third and fourth paradoxes mentioned previously. How can the need to understand be reconciled with the apparent impossibility of comprehension purported by the fourth paradox? The answer to this conundrum is favorable to our pursuit: what matters is the understanding of the extents and limits of the relationships, rather than the underlying workings of the technologies involved. Why?

The late Stephen Hawking (1942–2018) was a giant of modern physics. His work on black holes and singularities put him, rightfully, in the pantheon of the scientific immortals, alongside Brooks, Curie, Meitner, Noether, and Swan-Levitt, to name but a few.* His work spilled over many fields of esoteric research including anti-de Sitter spaces, large N cosmology, quantum entanglement, Yang–Mills instantons, Yang–Mills S matrix, Hoyle–Narlikar gravitation theory, and Brans–Dicke gravitation theory. These fields of research belong to a rarefied set of scientists and mathematicians, who struggle nonetheless to find even approximate theoretical solutions. We are talking beyond-bleeding-edge theoretical musings here. Incredibly, solutions may indeed be closer at hand than anyone could ever fathom just a decade ago, thanks to advances in computer chips, artificial intelligence, deep neural networks, and quantum computing. The most startling thing about these

---

* Harriet Brooks (1876–1933), Canadian. Physicist famed for her work on nuclear transmutations and radioactivity. Regarded by Ernest Rutherford (discoverer of the atomic nucleus, 1908 Nobel laureate) as second only to Marie Curie. Marie Curie (1867–1934), née Skłodowska. Polish. Nobel prize winner in physics (1903) and chemistry (1911): the only person ever to win two Nobel prizes in distinct disciplines. Lisa Meitner (1878–1968), Austrian. Physicist and discoverer of nuclear fission. Shamefully omitted from the Noel prize awarded to her research colleague Otto Hahn in 1944. Emily Noether (1882–1935), German. Mathematician who discovered the fundamental theorem on the connection between symmetry and conservation laws (without which the standard model of quantum physics could not exist). Considered by Einstein and others as the most important woman in the annals of mathematics. Henrietta Swan-Levitt (1868–1921), American. Astronomer and discover of the relationship between the luminosity and the period of Cepheid variable stars, which made it possible to estimate the distance to celestial objects. Without this law, Edwin Hubble would never have been able to quantify his discovery of the expansion of the universe (Hubble believed her deserving of a Nobel prize, again shamefully not granted). These ladies are but a few of the giants of science who are coeval with the likes of Newton, Faraday, Maxwell, Planck, Einstein, Feynman, Wiles, and Watson and Crick (another shameful episode of denial of contribution by a woman, Rosalind Franklin).

scientific and mathematical geniuses is that few of them can fully explain how the inner workings of the technological marvels support their efforts. For centuries, the theoretical brilliance usually implied deep mechanistic understanding of the machinery of experimentation. The bond began to break with the advent of quantum mechanics, when a division entered the realm between theorists and experimentalists. Nowadays, those working at the theoretical frontiers are satisfied to be users of the phenomenal tools at their disposal, without being their engineers as well. The increasingly cemented distinction between technology creators and technology users leads to the seventh digital paradox of the Symbiocene age:

> Paradox 7: It is no longer possible, nor desirable, for technology users to understand the engineering behind the inner workings of the digital tools at their disposal. What truly matters to these users is to understand the limitations of these tools, and how to interact with them to extract from them whatever insights and knowledge are possible.

This paradox is a gift from heaven to the multitudinous techno-heathens (including yours truly) who make up just about everyone else in this world. Someone must understand general relativity to make the global positioning system work properly; but nobody needs to have a clue about it to use a GPS device. For the time being at least, our world can learn to live and work with the technological prowess of the Symbiocene age, without any deeper knowledge of how that prowess is manifested.

# Bibliography

Babbage, Charles. *On the Economy of Machinery and Manufacturers* (original 1832). Reprinted by Cambridge University Press, Cambridge, UK, 2009, 320 pages.

Blanning, Tim. *The Pursuit of Glory – Europe 1648-1815*. Penguin Books Ltd, London, England, 2007, 708 pages.

Diamond, Jared. *Guns, Germs and Steel – The Fates of Societies*. W.W. Norton and Company, New York, NY, 1997, 480 pages.

Engels, Frederick. *The Condition of the Working Class in England* (original 1845). Reprinted by Pinnacle Press, 2017, 344 pages.

Freeman, Joshua B. *Behemoth – A History of the Factory and the Making of the Modern World*. W.W. Norton & Company, New York, NY, 2018, 427 pages.

Harris, Karen; Kinson, Austin; Schwedel, Andrew. *Labor 2030: The Collision of Demographics, Automation and Inequality.* http://bain.com/publications/articl es/labor-2030-the-collision-of-demographics-automation-and-inequality.aspx.

Keays, Steven J. *Investment-Centric Project Management: Advanced Strategies for Developing and Executing Successful Capital Projects.* J. Ross Publishing, Plantation, FL, 2017, 419 pages.

Keays, Steven J. *Investment-Centric Innovation Project Management: Winning the New Product Development Game.* J. Ross Publishing, Plantation, FL, 2018, 309 pages.

Sacolick, Isaac. *Driving Digital – The Leader's Guide to Business Transformation Through Technology.* American Management Association, 2017, 283 pages.

Schwab, Klaus. *The Fourth Industrial Revolution.* Crown Business, New York, NY, 2017, 192 pages.

Sills, Franklyn. *Foundations in Craniosacral Biodynamics, Volume One: The Breath of Life and Fundamental Skills.* North Atlantic Books, Berkeley, CA, 2016, 424 pages.

# Chapter 3

# The Nature of Data

*The best questions are formulated from the basis of informed igno-rance rather than enthused cluelessness.*

Information is like star quality: everyone knows what it is when they see it. Yet, few can define either without pointing to it. The result? Words likes data, record, information, knowledge, and intel-ligence are brandied about indiscriminately. Hence, the first step on the digital management journey is to get clarity on these terms, without which there can be no workable transformation strategy.

## The Spectrum

### *An Imperative, Not a Choice*

Whenever the topic of digital transformation comes up, one hears the admonition to establish a *data strategy*, which invariably conflates data and technology together. The attention earnestly shifts toward a focus on sexy applications, whizbang innovations, and miraculous computing capabilities. Pretty soon, the search for a data strategy morphs into a quest for cutting edge blockchain solutions, artificial intelligence operations, all-things-mobile, cloud deployment, and the ultimate prize of the platform effect. If the reader is like the great majority of business people interested in the subject, going digital being the mysterious "it" trend of the day, the prospect of undertak-ing the journey can appear wholly unsettling, like seeing a giant tsunami wave head your way. Fortunately, the reader need not face this wave at all. A profitable data strategy is about a firm's *data assets* and the means of turning them into investment vehicles. It is utter fallacy to believe that such a strategy can only be erected on a foundation of advanced, complex technologies sinews like AI, blockchains, big data, autonomous information processing, or cloud-based digital transactions. Indeed, in the vast majority of cases, the appropriate data strategy will involve at most a couple of tectonic threads discussed in Chapter 2, deployed across a revamped but *affordable* digital infrastructure which may not even be disruptive to the firm's day-to-day operations (see Figure 3.1).

All data strategies begin with *data*. The datum is the lubricant of any operation of a firm. It exists in all manners of volume, velocity, variety, veracity, and value[*] across the entire spectrum of human organizations. An organization like CERN, which operates the Large Hadron Collider (LHC), in Europe,[†] produces gargantuan amounts of data. A single experiment can yield 1 GB of raw data *per second*. Each year, the experiments yield 50 Petabytes—50 × $10^{15}$ bytes—of raw data, corresponding to 50 000 hard disks of 1 Terabyte ($10^{12}$ bytes) capacity each (or 10 million DVDs stacked 12 km high). These pale, nevertheless, in comparison to the astronomical scale of the internet. According to Waterford Technologies, a consultancy:[‡]

- The size of the internet is estimated at 2.7 Zetabytes (2.7 × $10^{21}$) in 2017
- BY 2020, over 450 billion business transactions will take place every day on the internet
- Walmart handles more than 1 million customer transactions every hour, which is imported into databases estimated to contain more than 2.5 petabytes of data

---

[*] As delineated by Bernard Marr, in *Data Strategy* (2017, Kogan Page). Volume refers to the quantity of data generated and transacted. Velocity refers to the speed at which a specific datum stream is generated and transported across information networks. Variety relates datum types and uses. Veracity alludes to the trustworthiness of data, defined in the widest possible sense in terms of reliability (can the datum be relied upon) and interpretation (is the datum complete, accurate, and legible if it is an abbreviation). Value pertains to the usefulness of a datum. Collecting data merely for the sake of it is utterly useless (and costly) if those data cannot be made to work for the business.

[†] The name CERN is a French acronym for "Conseil Européen pour la Recherche Nucléaire". The LHC is the world's largest and most powerful particle accelerator, where the Higgs Boson was experimentally confirmed in 2012. The byte is a unit of digital information that most commonly consists of eight bits (one bit has a value of either 1 or 0). One byte corresponded historically to the number of bits required to numerically encode a single character (a letter, a digit, a punctuation mark, etc.).

[‡] See www.waterfordtechnologies.com/big-data-interesting-facts/.

**Figure 3.1    The digital tsunami. The unceasing arrival of digital innovations into the marketplace can overwhelm the casual observer who stands on the shore, pondering how to get into the game. It is not possible to take it all in at once and make business sense out of it. It is always best to retreat far inland and wait to see which ones will be carried onward by the crashing wave. (Original artwork by Margaret Keays and Agnieszka Bak.)**

Clearly, the Symbiocene age is drowning in data, if not downright crushed by their relentless accretion at the planetary scale. Exploding data generation is also occurring at all scales, down to the smallest start-up company. Data are the life blood of a business', even to the most mundane of functions like accounting, inventories, taxation, payroll, and utilities. It is nigh impossible to find a *viable* business concern that does not handle digital data in one form or another, be they as Microsoft Office files or Google documents or emails or a basic accounting software. Even the humble barber shop or the one-man car repair shop cannot escape the imperative of dealing with data. Like it or not, your organization must accept the reality of the datum imperative. Resistance is futile; obduracy, fruitless. You have no say in the matter: data have already won this argument. But there is a wonderful upside to this defeat: data can be your friends, your decision-making advisors, your source of wealth. But data will be your enemies if you ignore them.

To know one's data is to know one's business.

## What Is Information?

The old saying "information is power" is incorrect. Information, in and out of itself, has neither power nor authority. One may possess a great secret without any way of acting upon it. That is not power... it is a tragedy. Power is the ability to act upon the information received. But the efficacy of that power to produce a desired outcome is also a function of the *maturity* of that information, i.e., its usefulness to decision-making? If you are given a temperature of 80°F, it means nothing. Add to it the context of location (say, Ko Olina in Hawaii), and you know that this a rather hot day. If it is Helsinki in January, you know that there is something terribly wrong with the weather there. If it's the temperature of your cup of tea, it may be too cold for your taste. Context, obviously, is critical. The numerical value is one characteristic of the usefulness of a datum. Usability is another; accessibility and authenticity are others. The realm of possible characteristics is in fact quite broad. It is necessary to cull it down to a limited operational set usable in a digital management transformation.

The resulting operational set comprises three subsets: the *knowledge spectrum*, for gauging the maturity of data for management decision-making; the *classification system*, for distinguishing the variegated types of data in terms of business value; and the *sentience spectrum*, for transacting data contents in various levels of digital autonomy. Note that neither subset takes into consideration the physiognomy of the data embodied into the medium (paper, analog recording, numeric file, etc.) and the format (handwritten, typed, filename.extension, electrical, 8-bit byte, audio file, video file, painting). Both are machine functions related to the capture, storage, retrieval and transmission of data which, in the aggregate, are subject to simple conversion processes from analog to numeric. The *contents* of data are not altered by medium and/or format, which is why they are excluded from our operational set considerations.

# The Operational Set

## *The Knowledge Spectrum*

This is the first subset, illustrated in Figure 3.2. The subset is used to assess the decisional value of a datum as a function of its *contents*. The spectrum illustrates the transformation journey of a datum from primitive instance into a strategic value. It can also be interpreted as the evolutionary path of an organization from an analog to a digital framework.

Information begins life as a primitive unit called *datum*. The *datum* is a simple fact, such as a number, a reading, a word, or an image. The *datum* is unstructured, unprocessed, and devoid of context. It may exist as a readout on a dial (say, the outside temperature), written on paper, spoken as an order, entered into a spreadsheet, or captured as a signal by a machine.

| Contents | Raw (as-is) | Structure added | Context added | Value added |
|---|---|---|---|---|
| **INFORMATION** | Datum | Record | Knowledge | Intelligence |
| **Value** | None | Token | Commercial | Strategic |
| **Maturity** | Primitive | Actionable | Reliable | Sentience |
| **Index** | 1 | 2 | 3 | 4 |

Figure 3.2   The knowledge spectrum. The "value" of information is a function of its contents, context, and capability to yield an assessment from which a decision can be made.

Its *maturity*—defined as the twin combination of usefulness to *actionable* decision-making AND machine accessibility—is equally primitive.

Structure gives the *datum* the ability to be transacted and manipulated, at which time it becomes a *record*. For example, the words "she left" forms a datum. The datum contains a primitive amount of information but not enough to do anything about it. They are unavailable to a machine when written on paper. Converting them into alphanumeric characters gives them a first level of structure. A second level of structure is provided by storing the converted datum into computer memory (or archive or whatever). We now have a *record* that can be processed at will, with a *maturity* raised to actionable (something can be done with that record).

> Note also that a datum differs from a record in terms of accessibility. If the information cannot be accessed other than by its creator, it is an orphan that is invisible to the organization (a sensor measurement that cannot be captured by a user, a file buried on the hard drive of an office worker, a record on a smartphone protected by password, a written form sitting in someone's truck, etc.). Any such information, regardless of its knowledge maturity, is a datum. Once access is endowed upon a datum, it becomes a record

Structure, however, is not enough. How does the meaning of these two words relate to the reader's circumstances? Was this person supposed to leave? Did she leave on time? Or was she expected to remain but left nevertheless? Was the departure a sad outcome (heartbreak) or a happy one (she was previously sick in a hospital)? It is amazing to consider how many interpretations can be derived from two simple words, and how many emotions can be ascribed to them as a result. Without context, the information meant to be conveyed by the two words is hidden. When the context is added to the record, the intended information becomes available. The addition of the context occurs when the record is created, or by processing the record via pertinent algorithms. The contextualized record is now transformed into *knowledge*. From this point forward, it is possible to decide what actions to take or not, pursuant to the information received. *Knowledge* is a record that can be relied upon. Maturity increases once again, from actionable to reliable. The record, as knowledge, stands on its own in the hands of humans or the commands of algorithms.

The further progression from *knowledge* requires one more level of processing whereby additional pieces of knowledge are integrated into a meta-knowledge called *intelligence*. *Intelligence* is the highest form of information, as it is the one that subtends the most potent form of decision-making power. Maturity reaches its highest level, one where the digitalized form of the intelligence can be imbued with *functional sentience*. Clearly, whenever it is possible, decisions should be made from *intelligence* rather than less mature information. Overall, the progression sequence across the spectrum is divided into four *knowledge indices*, as shown in Figure 3.2.

The "value" of information is directly proportional to its *contents*. The economic value of the *datum* is basically null, or at most equal to the cost of capturing it. The record only has a slightly higher economic value equal mainly to its processing costs. Its value can be higher if it is sold as part of a larger record set from which pattern inference can be performed. *Knowledge* information has a much higher economic value, as it can sold as actionable datasets (think online buying habits which generate customized digital ads to a specific user). *Knowledge sets* are commodities that can be bought and sold on that basis, with a sale price determined by the size of the sets. The king of information, *intelligence*, possesses an even higher economic value that may prevent it from being sold if it touches upon the strategic decisions of its owners (Google for example will NEVER sell its search algorithm or the cumulative knowledge derived from it).

Software plays a role at each stop along the information spectrum. Its complexity (and cost) increases in lock step with contents. Bits and pieces of software will be involved in capturing the data. Additional algorithms will be implicated in the structuring of the data into records. More complex instructions will be required to quantify the context to the relevant *knowledge pieces*, including AI in many cases. Industrial strength applications will be called for the processing of *intelligence sets*, most likely with powerful, deeply nested neural networks.

## The Classification System

The knowledge spectrum says nothing about the *nature* of the underlying data, which is required to set up the mechanics, mechanisms, and machine infrastructure to manage them. Understanding one's data therefore requires a different viewpoint through which the *value* of data can be quantified for business ends. An effective approach for doing so is through a *classification system*. The taxonomy introduced herein is explicitly limited to the commercial realm to avoid larger socio-political issues that are necessarily out of the firm's control. The classification system comprises five levels (mimicking the modern taxonomic scheme in biology): class, order, family, genus, and species, illustrated compactly in Table 3.1.

**Table 3.1  The Classification System**

| Class | Order | Family | Genus | Species |
|---|---|---|---|---|
| H | Exogenous | State | Prime | Datum |
| ... | | | | Record |
| 3 | | Cost | Derivative | Knowledge |
| 2 | Endogenous | | | |
| 1 | | Revenues | Adjunct | Intelligence |
| 0 | | | | |
| Level H: BoD | | | | |
| Level n-1: Leadership | | | | |
| Level 3: Management | | | | |
| Level 2: Management | | | | |
| Level 1: Management | | | | |
| Level 0: Line | | | | |

*Note:* The system borrows the taxonomy developed for describing the tree of life.

■ The *class* identifies the management levels of the organization, where each level is constituted of nodes (cf. Chapter 2). In the instance of a commercial firm, Class 0 corresponds to the lowest rung in the organizational hierarchy that generates data; Class 1 is the first reporting level for Class 0 nodes, then Class 2 in likewise fashion for Class 1 nodes, all the way up to the highest level of the organization, Class H (meaning the highest level), corresponding to the Board of Directors. Each node within a class can further identify what or who transacts the data (human, control, machine, or software) with a unique alphanumeric number in the form XZ, where X can be either H (human), C (control), M (machine), S (software), and Z is a one, two, three, or four digit sequential number (like 2, 02, 002, or 0002). In this manner, information is seen to flow horizontally between nodes of a given class, and vertically up or down between adjacent classes. The *class* provides the basis for assessing the valunomy* of the information flow within a firm's holobiome, through quantified efficiency, efficacy, warts,† and bottlenecks emerging from these information flows.

■ The *order* divides each class into exogenous and endogenous data, pursuant to the original partition of the holobiome agents. The endogenous order corresponds to data transacted between nodes of a given class, and between lower classes attached to a common upper class. The exogenous order corresponds to data originating from or intended for, a higher class. *In both cases, the order is established from the perspective of the node that transacts the data.* For example, say Class 3 corresponds to the engineering group, which is divided into five Class 2 functions (analysis, design, manufacturing, materials, and reliability). Furthermore, assume that Class 2 design is divided into four Class 1 disciplines (mechanical, electrical, chemical, and structural). Henceforth, all data streams flowing from the manager of the engineering group downward or to other coeval managers (say, procurement) will be endogenous *from the perspective of that manager* but will be exogenous *from the perspective of the lower classes receiving this information.* The *order* determines the *origin* or *source* of the data. It also identifies where control for the generation of the data lies.

■ The *family* defines the contribution of a datum associated with a state (report to a regulator, to shareholders, to any external entity), a cost, or a revenue. The cost family lies on the expense side of the general ledger, while the revenue family belongs to the profit side. The state family rolls both cost and revenue families for purposes of mandatory reporting, external compliance, or exogenous oversight. *Family* data drive profitability, capital investment options, innovation investment options, and financial management.

■ The *genus* divides a *family* into *prime, derivative,* and *adjunct* data. *Prime data* are generated for a specific, intended purpose. *Derivative data* are generated as secondary information associated (or derived from) process data. Adjunct data are generated for information purposes

---

* The neologism "valunomy" was introduced in *Investment-Centric Project Management*, Chapter 2. It is defined in opposition to the common expression "cost effectiveness". The usual definition of the latter is the least costly to buy now; whereas valunomy is defined in terms of the cost that will deliver the largest investment returns later.

† Refer to Chapter 1 for the definition of "wart".

only, without further business valunomy. For example, a thermostat will measure a room's temperature as *prime data*. The electricity consumed by the temperature control equipment will be reported on a monthly bill as *derivative data* associated with the temperature measurements. The name of the energy supplier appearing on the bill exemplifies an *adjunct datum*. The *order* determines the usefulness of data to the firm.

■ The *species* assigns to each datum of a genus a *knowledge index*, pursuant to Figure 3.2. The species is effectively both a measure of a datum's *value* and its *maturity*.

---

### CLADS AND KINGDOMS

Humanity has been obsessed with naming and classifying the world surrounding it since time immemorial. But a consistent, ruled-based approach to classifying the life forms in this world had to wait until the 18th century by the likes of Linnaeus (circa 1735), de Jussieu (circa 1789), Condolle (circa 1813), and Bentham and Hooker (circa 1862). These systems were unfortunately flawed by the fact that they came before Darwin's theory of natural selection (1859). The latter became the cornerstone of all modern classification schemes governed by evolutionary linkages (such as those proposed by Eichler in 1883 and Engler in 1886). The current classification basis takes into account molecular genetics to create cladistic "phylogenetic systems". The classification system comprises nine levels: life, domain, kingdom, phylum, class, order, family, genus, and species. You, the reader, are from the domain "eukariota", kingdom "animal", phylum "chordata", class "mammalia", family "Hominidae", genus "homo", and species "sapiens".

---

## Sentience Spectrum

The tire fabrication example in Chapter 2 was accompanied by the distributed network structure of Figure 2.2. Each node on the network represented a junction where data arrived as inputs manipulated by a local transformation process that produced one or more data output steams. We thus speak of a nodal configuration for the entire network, by which data are transmitted, transformed, and transacted by discrete nodes. What happens at a given node, at what level of data processing complexity, is defined by the *sentience spectrum*, which is used to determine the integration requirements of a node into an enterprise-wide digital network, and also define/select the desired level of digital autonomy to be vested into that node.

The nodal representation of Figure 2.2 is inherently fractal, which can be exploited to rapidly engineer an enterprise-wide digital transformation. Fractal means that the structure is replicable at all scales, whether the node is a simple device like a sensor or a valve, a system such as a metering station, an installation like a backup power system, a plant, or the network proper joining them together. The *nodal structure* in turn can take one of seven forms according to the complexity of the transformation process (of inputs into outputs). The general details of the structure are shown in Figure 3.3.

This structure can be applied to machines, humans, software, automated control systems, and a firm's management structure. The seven forms correspond to seven layers, each comprised of inputs, outputs, and the

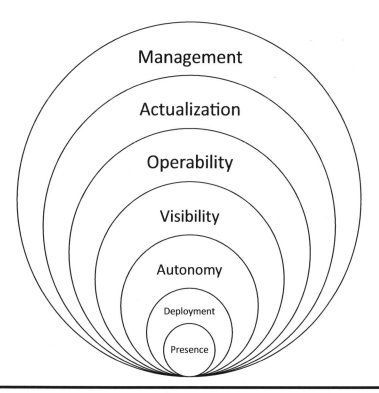

**Figure 3.3 The sentience structure. The structure illustrates the increasing complexity of a transformation process of inputs into outputs across seven layers. Because this structure is fractal, and therefore replicable at all scales, it can be defined once and applied to the simplest of nodes (like a single point sensor) to an entire installation.**

transformation process. This input-transformation-output scheme is also fractal and called the *unit transformation process* (UTP), described later in this chapter.

---

### A NEW DEFINITION ASSOCIATED WITH THE NODE

The Node is a symbolic representation of the unit transformation process (UTP) performed at that location to transform the data inputs arriving to it into data outputs to be transmitted outward across the network. This transformation could be performed by a single action sensor (say, temperature measurement), a mechanism (a pump pushing a liquid down a pipe), a process (transform the analog temperature input from Fahrenheit to Celsius), an algorithm (keep a running average of the temperature readouts on the pump's liquid output in segments of 15 minutes), or a control operation (stop the pump if the output temperature reaches 104°F).

We will call the unit transformation occurring at the node the "activity" of this node, which could entail one or more unit transformations happening locally at the node.

---

**Layer 1: *Presence*.** The innermost layer is the physical manifestation of the *purpose** of the node. The *presence* layer includes the physical, algorithmic, and inner control functions associated with the purpose. It also includes the node's interactions with its immediate nodal neighbors. The node can be static (a device bolted to a wall) or mobile (an inspector travelling to perform a calibration check). This kinematic state belongs to the first level of the structure called the *presence*.

**Layer 2: *Deployment*.** This layer surrounds the *presence* layer by encapsulating the physical and informational transactions between the node and those that provide it with external control. Control is exerted through signals that arrive and leave the node to govern the node's normal and abnormal operations. Control is also exerted by the physical inputs and outputs required by the node's actions, such as power supply, cooling fluids, effluents and waste products, emissions and noise, lubrication, etc. The ensemble of signals and physical inputs/outputs goes under the moniker "deployment flows".

---

* The term "purpose" follows the convention defined in Chapters 3 and 4 of *Investment-Centric Innovation Project Management*. The "purpose" of the device or algorithm associated with a given node is the reason it was bought by the organization to be used at that node.

**Layer 3: *Autonomy***. The *autonomy* layer is where the digital industrial paradigm appears first, by augmenting the node with the capability to self-assessment its performance, in real time. The *autonomy* layer renders the node *aware* of its presence and its effect on its nodal neighbors. The capability can include such functions as monitoring the performance limits; self-regulate when operational exceedances or variations are detected; determine the adequacy of deployment flows; re-route commands to bypass operating problems (backup and redundant capacity activation); and reduce performance to maintain system integrity.

**Layer 4: *Visibility***. This layer takes the monitored data produced by the lower layers and performs comprehensive assessments of performance, operability, and reliability metrics. The layer also produces probabilistic predictions on the root causes of metric excursions beyond their design ranges. The layer is heavily reliant on algorithms (AI or others). The layer also packages the outgoing information streams to be sent to the next layer (what data is to be sent where, in what format, at what frequency, and on what feedback basis).

**Layer 5: *Operability***. This is the first layer beyond the node proper. The layer receives the information streams from other nodes' *visibility* layers. The streams are captured in a live database. The information is aggregated by groups of related nodes and subjected to additional analyses for presentation, visualization (including *virtual reality* [VR] and *augmented reality* [AR]), and archiving. This layer effectively introduces the big data into the digital industrial's framework. For example, each external node in Figure 2.2 represents a supply chain partner (independent or not), set up with its own nodal structure and supplying the results of its metadata analyses to the main operator.

**Layer 6: *Actualization***. The high-level data mining and crunching occurs next at the *actualization* layer. The layer operates on data and meta-datasets alike via data mining, statistical analyses, deep learning neural networks, and other AI-type algorithms. Global operating, economic, and maintainability metrics are derived. Undercurrents, trends, and hidden correlations are uncovered. Simulations of what-if scenarios and their outcomes are performed. Intervention strategies are formulated, weighed, and estimated.

**Layer 7: *Management***. The final layer is called, appropriately, *management*. The knowledge sets and intelligence information produced by the *actualization* layer are evaluated by managers and executives. Decisions are made as to what must be changed, fixed, replaced, altered, abandoned, or re-designed (possibly with the help of additional AI tools). Action plans are quantified and implementation timelines are chosen. Anticipatory management actions can be formulated and deployed as decisions before control moves outside of management.

# The Transformation Toolbox

## *The Tricorder Kit*

The three operational subsets double as process analysis tools for the elaboration of a firm's digital strategy. This doubling is what gives the subsets their real value to any digital management transformation. The *knowledge spectrum* for instance is applied at the management level for deciding what decision is to be made at what maturity level. The allocation of decision "jurisdictions" (details in Chapter 4) across the organization's nodal structure helps define the inputs, transformation, and outputs of each node's *activity*, as well as the validation and verification requirements for each decision. The complexity of the *activity* in turn leads to the identification of the pertinent algorithms needed to execute the activity—in effect, pointing the analyst toward specific strains of AI, blockchain, energy lattice, IoT, and visualization implementations.

The *classification system* is the primary tool for analyzing the information flows across the firm's nodal network. In conjunction with the *Unit Transformation Process*, the classification system leads the analyst to drill down into the weeds of the network to uncover what data are produced for what reasons, at what rate and for whom, as well as what is missing—from which the firm's *information baseline* is derived, pursuant to Chapter 4. The classification system is especially critical in determining which datum streams are valunomic and which ones are not. Finally, one obtains the *analog opportunity cost* of the firm's status quo (also discussed in Chapter 4).

The *sentience spectrum* is used to specify the desired extent of a node's digitalization, based on the decision-making requirements established previously by the *knowledge spectrum* tool. When a node is already digitalized, the spectrum helps to establish the degree of autonomy to be imparted to the node, and re-wire the nodal network as a function of the imposed decision jurisdiction.

## *The Unit Transformation Process*

The concept of the UTP was introduced in *Investment-Centric Project Management.*\* It is illustrated in Figure 3.4 (appearing originally as Figure 4.3 in *Investment-Centric Project Management*). The UTP consists of an input stream which is subjected to one or more internal transformations to produce one or more output streams. The inputs and outputs are visible to an observer external to the UTP, but not the transformations, which remain locked inside the box. We say that the UTP maps a set of inputs into a corresponding set of outputs (either one-to-one, one-to-many, many-to-one, or many-to-many, in mathematical parlance), through the internal transformations.

On the left of the image are two additional inputs to the transformation: the *attributes* and the *targets*. The former defines the pre-defined features that must be embedded into the outputs (for example, if the output is a drawing, an attribute would be the drawing number). The latter quantify the execution of the *activity* (design power consumption, conversion rate, labor inputs, productivity).

Two more outputs are seen on the right of the image: *characteristics* and *metrics*. The *characteristics* are derived from the contents of the output (in our drawing example, a part count is a characteristics). The *metrics* are the actual execution performance, relative to input *targets* (computing time was 3.7 microsecond instead of 2.9).

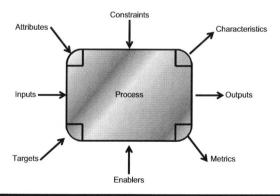

**Figure 3.4   The unit transformation process functionally. A node is an activity designed to map sets of inputs into sets of outputs, against pre-defined constraints and targets, and within the scope of enablers supplied by the network.**

---

\* The reader is referred to Chapter 4 of *Investment-Centric Project Management*, to explore further the idea of the *unit transformation process* (UTP).

The *constraints*, appearing on top of the activity, create the boundary within which the activity is executed (for example, a sensor cannot consume more than 24 VDC, the machine cannot exceed 67 decibels in noise level). The *enablers*, in opposition to the *constraints*, are supplied by the network to make the activity work (electricity supply, labor oversight, operating budget).

## *Direct Accountability*

In Chapter 5 of *Investment-Centric Project Management,* two fundamental principles of effective management were postulated:[*] 1) management occurs at the interfaces and 2) accountability is an *individual* mandate that comes into three flavors: accountability proper, responsibility, and authority, to which correspond, respectively, the *accountable party* (AP), the *responsible party* (RP), and the *probate party* (PP). The *first principle* states that a management intervention must only take place when an interface is in play, i.e., when the output of an activity is required to become an input to another activity. The *second principle* is encapsulated under the name *directrix* and stipulates that the AP can never be the PP for her own work. In short, the accountable party has the mandate to execute the activity (input-transformation-output) and produce the required outputs within the targets, constraints, and enablers supplied. The *approval* of the resulting outputs belongs exclusively to the probate party, which indicates that an output stream is permitted to be used as inputs to another activity. The probate party also defines the *constraints* for the activity. The responsible party supplies the attributes, targets, and enablers to the accountable party, and collects the characteristic and metrics.

The make-up of the *directrix* emerge organically from the structure of the UTP, as shown in Figure 3.5. When the two principles are applied to an organization, humans are in play in all aspects of the UTP. When they are applied to an information network, machines enter the fray, with algorithms or mechanisms playing the role of accountable party, and separate humans and/or algorithms and/or mechanics taking on the probate party role. The responsibility role, for its part, is built into the physical architecture (power distribution for example) and can also include more humans. Note that the prohibition against a single party holding accountable and probate mandates

---

[*] *Investment-Centric Project Management,* Chapter 5.

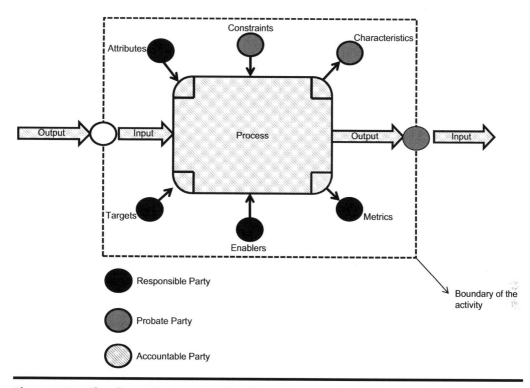

**Figure 3.5    The directrix at work. The directrix comprises the accountable party (AP), the responsible party (RP), and the probate party (PP). Each one is assigned specific roles in the execution of a UPT, without overlap.**

remains in place: even a machine needs independent approval of its outputs, if only because calibration drift will always be present.

# The Ugly Side of Data

## Second Cambrian Explosion

The advent of the personal computer in the late 1970s heralded a phantasmagorical explosion of forms of information. Numeric encoding techniques proliferated in complete chaos. Wars between competing file formats, operating systems, compression algorithms, audio-visual schemes, and image translations have been with us ever since. File formats have come and gone without warning or recovery plans. Simultaneously, the storage of numeric information proceeded along its own explosive abandon, thanks to the universality of standardized office applications. From the perspective of the historian or the archivist, the state of information dominion today is nothing short of a complete and utter catastrophe. On the one hand, too much information is stored in inaccessible ways. The constant churn in file extensions means that files created a mere decade ago have been orphaned on hard drives and portable media that are no longer readable. Readers old enough to remember the 1980s may recall the archaic 5" floppy disks and their more compact 3.5" brethren, neither of which is connectable to ANY modern computer. Or, the same readers may remember the glorious days of WordPerfect, CorelDraw!, or Lotus 1-2-3, none of which can be run on any computer today (and even if you could, the old files are still on those orphaned disks).

## THE MOTHER OF ALL GLOBALIZATION EVENTS

The history of life on the planet is, naturally, shrouded in mystery. Primitive life forms sporting soft tissues could have been preserved as fossils across time. What little evidence exists only gives tantalizing hints of what life looked like once earth settled into its nestling stage. Then, around 541 million years ago, everything changed. The fossil record from that era indicates an astonishing diversity of forms, skeletons, and structures. Thus, was born the Cambrian explosion, a unique period in earth's geological history that witnessed the sudden appearance of most major animal phyla (see the previous capsule *Clads and Kingdoms*). Life literally exploded into a carnival of new forms and new organisms of hitherto unknown complexity. Virtually all animal phyla in existence today appeared during this period.

This dreadful situation affects everyone and every organization on the planet. Even legendary NASA is struggling to come to grips with old data recordings made from leading edge 1960s technologies... Every computer user on this planet faces the same cancerous effects of antiquated legacy systems, orphaned file formats, incompatible recording hardware, obsolete operating systems, and anarchic file preservation processes. So much of a business' information holdings are kept randomly by individual workers in spreadsheets and documents, without any version control, content protection, or coordinated access privileges. This state of affairs leads us to the eighth paradox:

> Paradox 8: The visibility paradox plagues all organizations in proportion to their sizes: the more information transacts within an organization, the less visibility and control can be exerted by that organization over the information flow.

## Stranded Assets

In the extractive industries (mining, oil, and gas), the expression "stranded assets" refers to raw materials that are out of economic reach. The ore may be too deep or in too little concentration to justify the expense of bringing it to the surface. The oil may be plentiful but situated in a very tight formation beneath a deep seabed, making drilling exorbitantly expansive. Whatever the reason, the stranded asset represents a lost economic opportunity at best, or a wasted wealth pool at worst. Analogously, stranded assets in a digital information context represent the set of data, information, knowledge, and intelligence (together known as the *NAIA* [*numeric* and *analog information asset*]) design pieces that are unavailable to the firm, with the additional deleterious quality of being wasted sunk costs without any hope of cost recovery. Contrary to ore, coal, gas, and oil, this digital set had to be generated by the firm, then stored, manipulated, or archived, all of them incurring transactions costs to the firm.

> In effect, digital stranded assets are twice wasted and thrice expensed: the former when created and lost, the latter when created, then transacted, then as opportunity costs.

The importance of stranded assets to a digital management transformation cannot be overstated. Neither scale nor size nor market capitalization will spare organizations the bane of stranded assets. For the firm operating at the leading edge of information technology, the biggest threat comes from the inbred obsolescence of the volumes, contents, formats, operating systems (OS), machines, and algorithms used daily. The effect is the same for industrial control systems, plant automation processes, and computerized machinery (it is nigh impossible to upgrade a ten-year-old PLC or SCADA system used in a chemical plant, for example). This constant churn in the state of art inevitably causes a stagnation of the older art, followed by abandonment, and ultimately orphaned, at which point it is forever shut out of access and visibility. Anyone in possession of a WordPerfect novel written from a Microsoft Windows 3.1 OS and saved on a 5" floppy disk knows that this novel is utterly unreachable with today's capabilities.

At the other end of the spectrum lies 99% of all businesses and organization, for whom the biggest threat is increasing operating compatibility. Desktops and laptops and servers are usually late in upgrading to the latest OS or application releases. Most often, they will make do with what they have for as long as they can, allowing the obsolescence to grow slowly, invisibly until the day when the much-needed upgrades instantly cut off volumes of data and files and archives, accrued haphazardly over the years, from access and visibility. The picture is worse back in the shop, where little or no mechanical automation/robots exists, and where processes, procedures, and information transactions are performed by human labor, either on paper or archaic/primitive digital devices. Forget coherently comprehensive information management infrastructures; paper and static files (on someone's laptop) reign supreme, beyond any hope of governed control and oversight. The information created can still be voluminous but in a manner akin to flourishing weeds rather than tended gardens.

## Transformation Challenges

When dealing with information, one must do so under the aegis of the prime digital directive:

> Not all that can be captured or measured counts; but everything that counts must be captured, measured, AND accessed.

The subtext of this directive is straightforward. What matters to a digitalized organization is the ability to gather every piece of information with the potential for economic value (either measured, inferred, or endowed), and ignore the rest. It is this information that is subjected to constant real-time analysis to yield operational optimization (the point of the IoT, per Chapter 2). Eliminating the digital fluff, the useless, the repetitive, and the chaff also eliminates unwanted biases in the retained information, which could otherwise warp the deductive analytical results. Evidently, what is retained must be meaningfully quantifiable (the "measured" part) according to the analytical input requirements. And, it goes without saying, what is kept must be always visible, interpretable, and tamper-proof (the "accessed" part).

Here then is a summary of the challenges most likely to be encountered by the reader right from the outset, which must be addressed first before a proper digital transformation strategy can be enacted.

**Indiscriminate holdings**. Many organizations adopt as an operating guidance the principle of retaining all information created and transacted. This principle often arose from an absence of a well-defined information management policy rather than a prescribed directive. Furthermore, whatever and whomever creates information (a machine, a control system, an employee) is given free rein to create and store at will, without any checks or requirements to share these creations outside of one's own hard drive. The result is the modern office environment where each employee's computer is loaded at capacity with files and folders and emails, and where machine-generated data streams are captured unconditionally by servers and cloud storage solutions. Immense volumes of information keep accruing, without anyone in management aware of their contents, their costs, their merits, or their values. Such an environment is hopelessly useless as transformation candidates. Worse, it is destined to keep on making demands on the organization's operating budget without contributing one iota of a return on investment (ROI) to the bottom line. These information holdings can never be assets.

**Fluff and stuff**. This is a corollary of the indiscriminate holding. Without a sound information management policy, huge amounts of useless files and folders and datasets are retained, despite their contents being outdated, incomplete, erroneous, superseded, or obsolete. Think of all the draft versions of a single published document released by its author after multiple group reviews. Or, think of all the many versions of a file that exist simultaneously on the hard drive of everyone invited by email to provide comments (with additional versions kept in those emails as well). All this information is by nature useless to the firm yet incurring costs to continue to exist. And this is without even considering the data held in smartphones, tablets, and smart sensors.

**Insular holdings.** In this instance, disconnection is the issue. Even if the aggregate information was stripped of indiscriminate fluff, what remains would still be stored on myriad hard drives and servers devoid of any overarching integration and thus out of analytical reach. The organization ends up being a digital archipelago of thousands of datum islands going about their individual business without any connection to the others. Add in the mobile devices of the labor force and you get an idea of the impossibility of getting any kind of meaningful optimization results for the whole. Insular holdings are, in effect, the single largest impediment to any digital strategy.

**Illiterate files**. The expression "structured data" is used to describe a record located in a fixed field within a pre-defined structure of a file, with contents that are readily accessible and interpreted. In polite management circles, the expression "unstructured data" is used when data does not match the structured requirement. Pragmatically, the latter is illiterate, which means that a record can be found and even seen, but no sense can be made of it. For instance, NONE of the figures appearing in this book could be interpreted autonomously by an algorithm. Images, audio recordings, and video files are classic examples of illiterate files. Imagine for example that someone uploads into the firm's main public server 200 pictures taken during a field visit to a construction site of an airport. Then, a manager asks to see the pictures showing the progress of the water pipes running underneath the future landing strip. There will be no way for the computer to retrieve the pertinent photos without each file having been tagged with appropriate attributes by a person.*

**Innumerate contents**. This is the corollary of illiterate files, which plagues large datasets. Such datasets are typical of organizations who keep track of continuous transactions on a live database (daily sales, hourly production throughput, number of support calls per advisor, etc.). To be useful, the analysis results should be visually clear, which is not always the case when the database (or the spreadsheet) is only capable of primitive charts. Complex datasets may simply not be decipherable through statistical analysis (mean, average, deviation, correlation, etc.), or visualized with trend lines—and certainly impossible to do with static images when the number of variables exceeds 4. In this and other similar instances, the challenge is a disconnect between the hidden information in a valid dataset, and the machine capability to communicate this hidden information to a human or another algorithm. Even if the dataset is valid, valuable, and useful, it is innumerate to the outside world. The software is said to be divorced from its purpose (spreadsheets suffer especially from this divorce).

**Antiquated schemes**. This is the bane of every organization on the planet. The passage of time is unforgiving to data, software, and hardware schemes that inevitably fall into disuse from newer ones. Once they have been replaced, older schemes rapidly fall off everyone's radar, including the datasets created in accordance with their specific interface specifications. One needs look no further than paper archives accumulating dust in industrial storage facilities to get the point. At least these records can still be read by humans (unless the ink has faded or the paper has degraded). With numeric records, the losses are more dramatic—once they can't be read, they become useless.

**Invisible records**. Another unfortunate artefact of traditional information management practices is the disappearance of records, either from misfiling,

---

* Before suggesting the use of an AI application for the task, remember that someone will have to spend considerable time, effort, and money to "train" this application, which implies having access, beforehand, of an existing large set of pertinent pictures.

misnaming, or accidental deletions. In most cases, the record exists, but the user simply cannot locate it. The problem plagues humans far more than machines, of course. But when we include antiquated records that are physically obsolete (i.e., too old to be read by a modern device or incompatible with a current application), the problem of invisibility reaches humans and machines in equal measures. It may be possible to tally an inventory of those holdings, but such lists will always suffer from an inability to tie a line entry to the actual contents, since it is invisible. The bigger problem is that the archivist ultimately does not know the extent of what he does not know.

**Analog records**. For the 99% of the organizations pertinent to this book, analog records continue to be an essential component of their execution philosophy. Paper forms, legal documents, stamped drawings, video surveillance recordings, audio recordings, marketing materials, business cards: the list goes on and on. Paper in particular will never go away completely; there are some processes that are simply too efficient in that medium to be digitalized. Nevertheless, the proportion of records that must be analog is far smaller than the numbers currently held, which would be better served by a digital conversion. The challenge in this case is the hard habit to break; people can be very defensive of their analog habits, regardless of the cost to the firm. Unfortunately, as long as records are kept in the analog domain, they will be unavailable to the digital domain where real-time optimization takes place.

**Integrity**. This is the last universal challenge in our summary. Integrity in the digital context relates to the origin of a record and its continuation in time without alteration. Analog records lose their integrity primarily through physical degradation. Tampering with an analog record is much harder to do because it implies modifying the physicality of the recording (erasing something, forging a signature, altering the text). Analog records are also prone to theft (which is much harder to detect when dealing with archives) and misplacement (stored on the wrong shelf). Digital records do not degrade—but the medium holding them does. They can be accidentally altered by magnetic fields or by human action (much more prominent). The biggest issue with digital records is of course hacking, which is at heart an issue of security and recovery. The issue is at the top of most chief information officers' priority lists.

## Moving on to the Next Step

This chapter has opened a window into the nature of the beast to be tamed by a digital management transformation. The window was necessarily narrow because the topic is far too broad to circumscribe within the scope of this book. The intent of the discussion was to bring forth the major aspects of the business of data and information. Each reader will no doubt have to contend with unique circumstances not addressed herein; hopefully, the text will have afforded enough leeway to derive appropriate conclusions in your case. This chapter articulated an understanding of what data means to an organization and its bottom line. In the next chapter, the text will take the reader into the initial stage of a digital management transformation, by which the current status quo of the reader's organization will be compiled and quantified. This first stage is a necessary prelude to following chapter, where the mechanics of engineering a digital strategy will be discussed.

## Bibliography

Keays, Steven J. *Investment-Centric Project Management: Advanced Strategies for Developing and Executing Successful Capital Projects.* J. Ross Publishing, Plantation, FL, 2017, 419 pages.
Mar, Bernard. *Data Strategy: How to Profit from a World of Big Data, Analytics and the Internet of Things.* Kogan Page Limited, London, 2017, 200 pages.

# THE JOURNEY

**II**

# Chapter 4

# Data Demands

*The hidden lives of data: invisible actors existing inside networks within ethereal worlds.*

A digital management transformation begins with a thorough understanding of a firm's digital assets, stripped of technological considerations. The generation, acquisition, and usage of a firm's data is mapped into a conceptual framework from which valunomy and monetization can be engineered afterwards.

## Why Undertake the Transformation?

### The Meaning of a Strategy

This chapter engages the transformation process through the development of a *binary strategy*. We need to be crystal clear from the outset about the reasons why such an endeavor should be undertaken, starting with the meaning of the term "strategy". The word is ubiquitous in business, management, and academic circles, to the point where its meaning is tacitly assumed known by everyone. But when people are asked for the differences between "vision", "mission", and "strategy", or between "strategy, "tactics", and "tasks", the sense of certitude vanishes. Therefore, let us posit the following definition:[*]

---

[*] This definition is borrowed from the one stated in Chapter 8 of *Investment-Centric Innovation Project Management*. Other related terms such as vision, mission, value proposition, tactics, and tasks are also addressed in Chapter 8.

A strategy is a detailed plan to orchestrate an organization's resources, budgets, timelines, and milestones to realize a given mission—which is, in our context, to transform the organization into a binary firm. The strategy breaks the mission (the endeavor to put the strategy into action) down into discrete work units, then sets out the targets and the metrics to achieve them. The strategy defines up front what is to be included and excluded from the endeavor. It also defines the overall hierarchy, the decision-making principles, the conflict resolution mechanics, and the accountability assignments. To be effective, a strategy must be realistic; and to be realistic, it must intelligently exploit the strengths of the organization and neuter its weaknesses.

Pursuant to this definition, the digital management transformation is to be governed by the following strategic goals:

- **The mission** of the binary strategy is to transform the organization into a binary firm able to coalesce its analog and digital assets into a seamless operating principle to achieve higher profitability, efficiencies, and shareholders' returns on investment over the long run.
- **The purpose** is the elaboration of an enterprise-wide plan to engineer the *digital construct*, design the analog-digital interfaces, implement the transformation elements, and deploy the newly formed binary operational concept across the organization's activities *without breaking the bank*.

## Binary vs Data Strategy

Why use the expression "binary strategy" instead of "data strategy", so popular in management consulting circles? Advocates of the expression "data strategy" militate about the urgency for organizations to get on the digital bandwagon, regardless of their unique circumstances. They put data squarely at the heart of everything and insist on deploying complex algorithms and costly hardware infrastructure to massage those data in real time, all the time. These exhortations are pertinent to a fraction of firms and organizations driven by mobile customers, ephemeral trends, evolving tastes, or far ranging logistics. But these exigencies are at most incidental to the great majority of the remaining firms and organizations to whom this book applies. There is no life-or-death urgency to become a software company, to deploy sophisticated artificial intelligence software to analyze a firm's spreadsheets, or to blockchain every business transaction just because it's the "it" thing to do at the moment. The urgency lies elsewhere, in the necessity of recognizing the fact the world has gone digital. But there is more to going to digital than focusing on the digital construct. Contrary to "data strategy" proselytes, who see the digital construct as the end unto itself, a binary strategy regards the construct simply as a means to an end, which is to enhance a firm's profitability, efficiencies, and returns on investment to its shareholders.

> In a digital management transformation, the mechanics and mechanisms of the digital construct are chosen as a function of the decision-making requirements of the organization. The digital construct FOLLOWS the requirements, never the reverse.

This statement is critical to understanding the transformation methodology. The bits and pieces that go into the digitalization of the organization (bought off-the-shelf, customized, or bespoke) are *irrelevant* initially in the grand scheme of things. Their importance materializes only when it is time to integrate them into the workings of the organization. Indeed, in many instances, whatever is on hand will suffice without further ado or modifications.

## *The Sequence of Events*

The binary strategy is divided into four parts carried out sequentially. First, one must understand the state of the organization's information assets today, which will often begin with little or no grasp of exactly what is in play, why it is in play, how it is working, and how it is failing. This is the essence of Part I of the binary strategy, explored in this chapter. Completing this important first step is essential to the success of the second part. It provides management with precious insights into the inner workings of the organization (explicit and implicit). Armed with the knowledge of one's status quo (the great, the good, the bad, the ugly, and the scary), management initiates Part II (the subject of Chapter 5). The purpose of Part II is to envision what can be done with the assets, given their known limitations and potential capabilities. The onus for Part II is on the organization's leadership, who are tasked to imagine how the assets can be monetized conceptually. The imagining process relies on a specific methodology for deciding how to best exploit the monetization potential of those information assets. Part II yields a set of strategic targets regarding the desired valunomy to be achieved from the revamped information assets. It is essential to the success of Part II that technological considerations be left off the table, lest preconceptions and misunderstandings place unnecessary or harmful limits upon the targets. The technological considerations are taken up by Part III, through which practical (not theoretical or bleeding edge) solutions are assessed for suitability to manifest the targets of Part II. Part III yields a complete set of functional and performance specifications for the hardware, software, and processes that will form the working components of the new digital construct. Part IV pulls together the findings of Parts I, II, and III by formulating a binary framework comprising the corporate binary policy, the policy execution plan, the implementation of the pilot digital construct, and the full roll-out of the binary framework throughout the organization.

> Together, the four parts form the overall strategy development road map, which acts as the execution plan for the digital management transformation.

## The Transformation Team

The digital transformation requires a dedicated team. The team is headed by the transformation manager (or chief or director, according to the organization's habits). This transformation leader must report directly to someone within the executive team (preferably the CIO or the COO in larger organizations, the CEO or the founder for smaller ones). This line of sight to the C-suite is essential to the success of the journey: without a mandate from the leadership of the organization, failure is likely. The make-up of the team will comprise a number of crews assigned as a function of the skillsets needed by the work at hand. Part I will involve a *discovery crew* and possibly an *audit crew* (see below). Part II will involve a *target crew* whose task will be to quantify the digital transformation targets. Part III will switch to a *technology crew* augmented by consultants and assisted by the IT group to define, assess, and select the hardware, software, and system performance requirements of the solutions needed to give form to the targets of Part II. Finally, Part IV will involve four crews: one for designing the performance and integration specifications of the *digital construct*, one to define and implement the *governance policies* of the transformed organization, one to plan and implement the pilot transformation, and one to plan and roll-out the implementation to the rest of the organization. The IT group will be involved as well in the work of each of the five groups. Schematically, the team structure is shown in Figure 4.1.

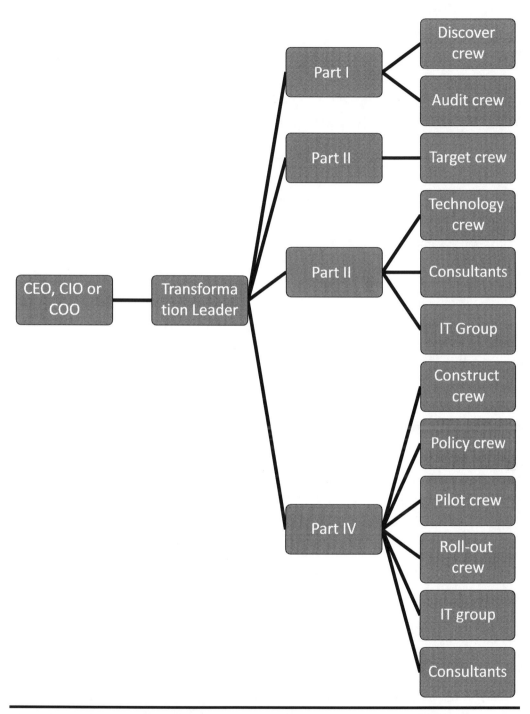

**Figure 4.1    The transformation team. The functions of each crew are defined in terms of the skillsets needed by the tasks of each part.**

# Baseline

## *The First Milestone*

The first step of all digital management transformation journeys is to establish a baseline of the organization's digital assets. The baseline is the first milestone of Part I; it establishes the limits of the assets' *status quo*, which informs management on *what is, what must be,* and *what is not.* Qualifying this *status quo* is fundamentally a process of discovery, with the emphasis on the information *contents* as they exist today. Things like formats, media, applications, and hardware are captured as well, but are secondary in importance (since these could change during the design of the digital construct).

The general mechanics of discovery is fractal, i.e., it works at all scales of the organization's hierarchy, using the *classification system* introduced in Chapter 3 (see Table 3.1) as the primary tool. For a given class, the nodes in that class are selected for analysis when data flow through them to be transformed from inputs to outputs *for the purpose of decision-making.* This caveat means that not all nodes will be retained for analysis, nor that all datum streams will be captured. An office worker may, for example, be receiving daily reports and extract from them specific data entries to be gathered on a separate spreadsheet used only by that worker. Although information flows to this worker-as-network-node, the transformation of the input data into an internal record (the spreadsheet) does not lead to a decision. Hence, neither the node nor the datum stream is applicable to the baseline analysis.

The discovery process proceeds for all nodes within a class (including all lower classes within a given class), then across horizontal classes, then to and from vertical classes. The scope of the discovery can be limited, at the outset, to a single class grouping, then be extended over time to other class groupings to better manage the level of effort and reduce the risks to on-going operations. The extents of the discovery work must abide by the prime digital directive enunciated in Chapter 3:

> Not all that can be captured or measured counts; but everything that counts must be captured, measured, AND accessed.

The discovery is not intended as a comprehensive audit of every possible morsel of information generated whatever the reason, whenever the case.

Such an extent is neither valunomic nor useful: there is simply too much information floating around. The point of the discovery is to sort the wheat from the chaff and focus on what is essential to decision-making and monetization. This sorting is accomplished through the application of the classification system to every input and output of every node, with the most important classification qualifier being the *genus* (prime, derivative, adjunct) at this time. For it is the genus that will enable the discovery crew to determine what is essential (the prime), what is necessary and possibly valunomic (the derivative), and what can be ignored (the adjunct).

No attempt is made during the discovery to address issues other than contents (things like format, file types, applications, storage, etc.). These features belong to the *analysis* stage of the binary strategy. No attempt is made either on deciding what are the best means of creating the data. In other words, if a node is currently an individual who produces, by hand, a weekly tracking report on late purchase deliveries, then so be it for now (as opposed to insisting that such a report could be automatically generated by the organization's ERP* system). On the other hand, the transaction identifier (described in Chapter 3 as the alphanumeric number XZ within a Class) is an important characteristic to capture for a node uncovered during *discovery.*

The discovery work will yield a large database of findings—*indeed captured in a database, not a spreadsheet*—from which a preliminary version of the baseline can be outlined. It is IMPERATIVE that the findings be recorded in a database and not a gaggle of individual spreadsheets created by the members of the discovery crew. The follow-on analysis work *simply cannot be done effectively from scattered spreadsheets or other kinds of tabulated documents.* Recall the insistence on *access* by the prime digital directive, which can only be satisfied by a database or another similar record capture system (including the Cloud). Maintain the numbering scheme derived from the classification system as the basis of identification for the findings captured as records in the database.

The final component of the discovery mechanics is a compilation of the findings in various forms including:

---

* ERP stands for Enterprise Requirements Planning. ERP systems are typically employed to centralize all financial and accounting activities of an organization. Often, the range of activities will be expanded to include activities associated with procurement, human resources, inventory and warehousing, labor tracking, project management, and construction cost control, to name but a few.

- The first complete picture of the information network of the organization
- A detailed listing of the actual make-up of the nodal classification of the organization (how many classes, how many subclasses for each class grouping)
- A quantification of the nature of the datum streams and information flows transacting at each node (i.e., the full classification of each input and output of each node by class, order, family, genus, and species)
- The first draft of the lists of "what is" and "what must be"
- The initial assessment of "what is not"
- The first glimpse of the warts pullulating across the organization

The sum of these compilation forms adds up to the organization's *digital asset* baseline. This baseline can now be used to figure out what must be done, what can be done, and what should be done with these assets. "What must be done" aims at fixing the existing warts and obliterate the value-destroying features of the status quo. To fix the warts, optimize the information flows, extract more valunomy out of the status quo, and implement immediately whatever remedies are needed to meet the unmet obligations of the organization (pertinent to the "what must be" information flows). "What can be done" speaks to the existing capabilities and limitations of the digital assets as well as the underlying digital construct in place at the moment. It is a snapshot of the way the organization creates, manages, and transacts information, along with the technical means of doing so.

## Getting to "What Is"

One can look upon the discovery process as an audit of an organization's digital assets, without any preconceptions of what will be discovered. Much like its accounting kin, the audit is hands-on and labor-intensive. The audit will be as successful as the extents of the access offered to it. The audit is conducted either by the discovery crew or a designated audit crew, tasked to ask a wide variety of questions:

- What are the inputs coming to a node?
- What are the outputs leaving it?
- What are the physiological features (format, contents, structure, limits, rates, etc.) of those inputs and outputs?
- Which information is analog? Which is digital?

- Which analog information can be digitized? Which cannot be without significant investment/re-structure?
- Who/what needs what information, when and why?
- How is the information stored/archived/updated/maintained/version controlled?
- What information should be version controlled but is not?
- What information is invisible, orphaned, or siloed?
- Who/what has access to the stored information, with what permissions?
- How are read/write permissions managed, by whom/what?
- Are records updated in accordance with a formal management policy?
- Who/what owns the records pertinent to the node?
- What measures are in place at the node to prevent hacking and tampering?
- How are the inputs, outputs, and records verified for correctness, veracity, and legitimacy?
- What information is NOT recorded by the node, but could be?
- What information is NOT recorded by the node, but should be?
- What information is NOT recorded by the node, but must be?
- What are the UTP constraints externally imposed on the node?
- Which information belongs to the exogenous order? To the endogenous order?
- Which information is not transacted outside of the node?
- What nodes furnish the inputs to the node under review?
- What nodes receive outputs from the node under review?

This question sample covers the essential areas of enquiries but is by no means complete. Other questions will arise from the unique circumstances surrounding a specific node. The bottom line of the audit is this: whatever question is necessary to quantify the information existing at a node is an admissible question.

Once again, the reader is cautioned against applying the audit questions indiscriminately to all nodes, pursuant to the prime digital directive. *The information is pertinent to the binary strategy if and only if it is involved in decision-making*. Without an assignable role to decision-making, the datum/record/information/knowledge can be ignored, for the time being, by the binary strategy. Therefore, if a piece of information is clearly superfluous, futile, superseded, expired, or otherwise unused within and without the node, there is no need to include that information in the audit. This will be the case most obviously for the files, folders, and documents existing on the hard drives of workers' laptops and computers.

## *Getting to "What Is Not"*

Finding out "what is" only requires asking open-ended questions, and recording the answers as they are, unfiltered. Getting to "what is not", however, requires an intermediate step between the two, one that assesses the ramifications of the answer. For example, a retail store may be keeping track of the time of day when sales occur, one which day of the week. The audit will uncover this finding and file it as a "what is" record. But what about the larger context in which the tracking takes place? That is the essence of finding out "what is not". One goes beyond the immediate answer and attempts to determine what other factors could be influencing the answers but are unknown by the organization. In our store example, one could ask why they are not keeping tabs of sale events by direct competitors inside the mall and by all other stores in the mall? Or, ask whether or not these external events correlate to variations in the in-store metrics. Or, ask what sales are lost by the store on account of a different offering by a competitor? During the discovery stage, the objective of the "what is not" enquiries is to identify those questions for which an answer is obviously missing at this time.

Other "what is not" questions will address the observed limitations of the status quo. For example, assume that a refrigeration process monitors the inside temperature and generates a continuous analog signal back to a control panel. The panel converts the signal to a digital value and records the datum stream as discrete values in one-minute intervals on a hard drive. This is the "what is" picture on hand. The "what is not" assessment could ask why the outside temperature is not recorded in tandem? Why is the inside temperature not transmitted as a digital signal in one-minute intervals? Why is the control of the internal temperature performed remotely, by the control panel, rather than *in situ*, with a closed loop controller driven by the inside temperature? And why is the inside temperature pegged at whatever value it is now? Could it be raised without deleterious effects to what is kept cold? Why is the inside humidity not recorded as well? Alternatively, the "what is not" questions could be applied to determine is information not actively recorded at a node. Let's assume that the refrigeration process is equipped with an internal thermometer that is connected to a close loop *in situ* controller that turns the cooling system on or off as a function of the internal thermometer's reading. The "what is not" questions in this case would simply be inverted. For example, one could ask why this internal temperature is not monitored by an external control system tied into the shop's overall power monitoring system.

Yet another group of "what is not" questions will deal with the limitations of the status quo *in relation to the external features of the organization's holobiome.* A business, for instance, could be operating in a rigid regulatory structure that obligates the business to produce monthly reports, for which data is not currently gathered by the business (such as the mandatory reporting of greenhouse gases emitted from operations, to give but one example). While the business is aware of the requirement, it could be that its digital construct status quo is not set up to capture this information. Another external holobiome feature could be the practice of a competitor to monitor, in real time, the whereabouts of its fleet of service trucks, subjected to statistical analysis to minimize travel time and/or distance (think FedEx and UPS for example). This competitor will eat the business' lunch in short order if the latter is not capable of such monitoring and analysis. At an even more basic level, it could be an instance of a routine business process that is NOT producing an information stream on its performance metrics, which would be highly useful to management (think personnel productivity in an office environment).

One can regard the "what is not" enquiry in terms of a SWOT analysis (strength, weakness, opportunity, threat) familiar to business process engineering. Under this guise, the "what is not" questions will concentrate on the weaknesses and the threats, with such questions as:

- How and why do you generate your datum streams?
- How are these streams flowing and where?
- What are the external data demands?
- How are they trending?
- Which data streams are cost drivers?
- Which data streams are profit makers?
- What is the competition doing with their data?
- Is the competition dealing with datasets that are different/new from yours? Why?
- What is the impact of these different datasets on your efficiencies, your productivity, your profitability?
- Who is eating your data lunch?
- What can you NOT provide today that is externally expected/requested/ mandated?
- How soon must you overcome it?
- Where do you stand, in information terms, within someone else's supply chain?

- Where do you stand within someone else's regulatory scheme (think GDPR*)?
- Which jurisdictions are a threat to your intellectual property?
- How safe is your digital construct? How hackable is it?
- What new data requirements could arise in the short, medium, and long term that you can't deliver, which can or could be material to your bottom line?

## Getting to "What Must Be"

The third question set of the discovery takes in the answers from the "what is" and the "what is not" questions and extracts from them the information requirements† that must be satisfied NOW by the organization, regardless of the limitations of the status quo. Simply put, the organization has no choice but to comply with these requirements lest it suffers deleterious impacts to its bottom line or runs afoul of regulatory predators. Some of these requirements will be obvious, such as those mandated by law, code, regulation, or contracts: a public company is governed by strict disclosure requirements; a large company may be operating in a jurisdiction with an enforceable diversity mandate; a registered charity must file an annual financial report to the tax man. Others will be internally driven to enable management to make business decisions (weekly sales reports, staff turnover, internal rate of return, quarterly EBIDTA metrics, etc.). Contract requirements are obviously mandatory to all parties who are signatory to them. Operational requirements are another class of unconditional information demands (whatever is needed to keep the production line running smoothly, for instance). The variety of non-discretionary requirements is endless but always a function of the internal and external circumstances of the organization's holobiome.

> The exogenous "what must be" information requirements define the relationships of an organization within itself and within its holobiome, with the latter acting as an accelerator of competitive pressures.

---

* GDPR: General Data Protection Regulation. This regulation, with the force of law across the European Union, came into effect in 2018 to protect an individual's data and privacy.

† A *requirement* formulates a *need* in terms of what, why, and when. A *specification* quantifies the *need* in terms of who, where, and how—how it is realized, delivered, and operated. For example, people need money. Money is a requirement. How much money, on the other hand, is a specification.

The last part of the above sentence is crucial to grasp the importance of a digital management transformation. The primary source of threatening pressure to any organization, in the Symbiocene age, is the permanent state of advancement in information technology on the outside, which confers upon its adopters an immediate competitive advantage. Digital advancements continuously arise and suffuse one's holobiome, which in turn create or impose compliance exigencies that cannot be ignored. Hence, it behooves all organizations, regardless of their financial model, to remain alert to the evolving character of their holobiome's "what must be" requirements.

There is a difference between what must be and what can or could be, and it is crucial to understand the implications. The former deals with information in its essence, without paying attention to the handling mechanisms (the software, the hardware, the transmission channels, etc.). The focus is on what is required rather than how it can be realized, which speaks to what can be or what could be. Take for example a support hotline to assist customers. The mandatory requirement is for the hotline to provide answers to incoming questions. That is the "what must be". Realizing this requirement can be done by a person, a—dreaded—recorded message, or a fancy human-like machine interface running on artificial intelligence bits. All of these options, and perhaps more, *could* be deployed, but it does not mean that they *must* be so, especially when considering costs. One can easily become enamored with the glitz and glamor of fancy algorithms that bring leading or bleeding edge technology to the organization. The danger lies in falling to identify what is best for the realization of the requirement. In our example, the better solution will always be a real person to answer the phone. The individual might be assisted in the task with one or more fancy solutions, running in the background. But the human connection in this case prevails. This example illustrates the importance of separating the requirement (what must be) from the specification (how it is realized). It also helps to avoid the pitfalls of confusing whims and necessities.

# The Analog Opportunity Cost

## *Parsing the Baseline*

Together, *what is, what is not, and what must be* constitute the organization's baseline, as well as the state of its information status quo. It is a snapshot in time of the capabilities and limitations of one's information assets. The next step in the audit process is to parse these capabilities and limitations through the "can–could" analysis, comprising three stages:

■ Stage 1 "what can be" assesses what additional value can be extracted from the status quo through organic modifications. These modifications could entail software/hardware upgrades, new application deployment, re-design of business processes and procedures, or organizational re-structuring. Stage 1 yields a new and improved status quo.

■ Stage 2 "what could be" goes beyond the status quo by determining what else is available to the organization in terms of efficiencies, pro-ductivities, revenues, and profitability when the status quo is left out of considerations. "What if" scenarios are explored for their potential to deliver upsides against a backdrop of associated costs, operational disruptions, and corporate upheavals. Changes are likely going to entail more dramatic impacts to the status quo, and definitely to the short-term bottom line. But these changes also carry with them the poten-tial for much greater commercial benefits to the organization over the medium and long terms. Stage 2 can radically transform the status quo into something entirely different.

■ Stage 3 "opportunity costs" reveals the missed opportunities and block-ages inherent to everything that is analog in the organization, and to poorly designed digital processes. In effect, Stage 3 tells manage-ment why the desirable elements of Stages 1 and 2 cannot be pur-sued because the status quo is stuck in the mud. Part 3 also informs management of the inefficiencies and value-destroying aspects of the status quo. For example, if employees are required to submit a weekly timesheet written on paper, or via individual spreadsheets, the analog opportunity cost is in the unproductive labor required to manually enter the timesheet data relative to the instantaneous compilation that would be performed by a fully digitalized database application integrated into the firm's ERP system.

## The Second Milestone

Taken together, the three parts of the analysis amount to a complete picture of the organization's analog opportunity costs. The aggregate benefits form all that can and could be are simply denied to the organization because its information status quo stands in the way. The organization may be satisfied with how things are running now, perhaps even going so far as to refuse to change because of a belief in the saying *"if it ain't broke, don't fix it"*. The status quo reigns supreme and deludes management into thinking that all is well. Until, that is, management looks outside the front door and sees the competition enjoying a sudden spike in gross margins because of a successfully implemented binary strategy. The point should be obvious to the reader concerned with competitiveness: in an age of relentless improvements in information technology, commercial oblivion awaits the laggard who clings to his status quo. That status quo is measured by the foresworn profitability carried by its analog opportunity cost.

## Planning the Transformation

### *An Example*

We consider a fictitious commercial firm called Industrial Fabrication Ltd, or IFL for short, whose business is to design and fabricate specialized industrial trailers hauled by road or rail. The trailers are made principally of welded steel elements and come equipped with mechanical systems such as rotating equipment, pressured fluid storage, process monitoring, and space heaters. IFL has its headquarters in Lafayette, Louisiana (USA), and operates two fabrication subsidiaries in Neufchâtel, Québec (Canada) and Xalapa, Veracruz (Mexico). The fabrication operations are overwhelmingly manual, without robots or automation, and are governed by an ISO 9001 quality assurance system* relying mainly data capture by humans. The corporate hierarchy is shown in Figure 4.2 and that of the Neufchâtel division (where French is the primary language) is shown in Figure 4.3. The application of the classification system yields the following details of the information network:

■ Figure 4.2 comprises seven class levels, from 0 to 6, with 6 assigned to the Board of Directors.

---

* Notwithstanding this statement, one can go through the motions of mapping out the digital network of the present status quo. The exercise is akin to a traditional business process analysis to reveal who is involved in what internal transactions of the organization. The value of the exercise rests upon the possibility of unearthing the value-destroying redundancies, inefficiencies, and bottlenecks that lie hidden beneath the management surface. First timers involved in the process often come away dumbfounded by the extent to which these warts have accumulated over the years without anyone noticing them. Such an analysis is *always* beneficial to the organization's leadership but comes at the risk of exposing the ugly sides of power grabs, silo chiefdoms, and self-interest by jealous managers. As the old adage goes, *if you are afraid of the answer, don't ask the question…*

- The corporate suite sits at class level 5, with a single instance assigned to the CEO.
- Two senior leadership positions sit at class level 4. The designation n1 and n2 uniquely identifies these direct reports to the CEO.
- The VP of Manufacturing (4n1) oversees the three geographic divisions (Lafayette, Neufchâtel, and Xalapa), as well as a planning group, representing the third class level for 4n1, numbered respectively 3n1, 3n2, 3n3, and 3n4.
- The VP of Operations likewise oversees four operational groups numbered 3n1 through 3n4 as well.
- Figure 4.3 dives further down into the information weeds, where we find, under the Warehouse Foreman (class level 2, number 1: 2n1), two additional class levels, 1 and 0.

Note at this point that the nature of the information generated at any class level is not captured. The numbering scheme exhibited in both figures is limited to the number of classes present in the organization (cf. Chapter 3). Defining this class stratification is in fact performed as the first step of the discovery process. Once the stratification is on hand, the actual identification

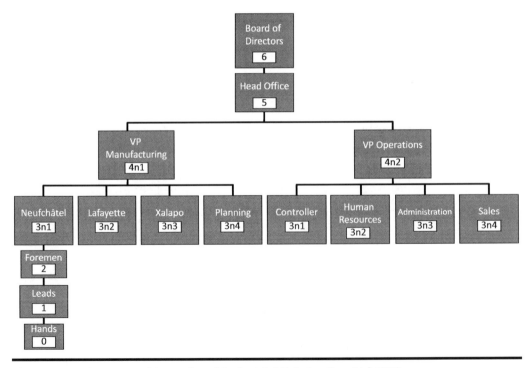

**Figure 4.2   Corporate hierarchy of Industrial Fabrication Ltd (IFL).**

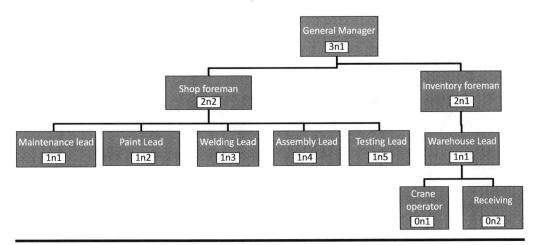

**Figure 4.3    Reporting structure of the Neufchâtel division.**

of the data and information transactions can commence. Focusing now on the Receiving function 0n2 under the warehouse lead (1n1) reporting to Inventory Foreman (2n1), the following nodes would be discovered:

- Materiel shipped by a vendor is accompanied by a bill of lading in printed form. Since information is transacted (received), the receiving party becomes a NODE which, in this case, is transacted through human action, and hence identified as a type H node, then numbered as Node 1H01 (where the 01 implies this is the first individual among a group of people associated directly with this node).
- The bill of lading is an information input classified: Order: exogenous; Family: cost; Genus: prime; Species: information.
- The bill of lading is visually checked by the receiver against the original purchase order and noted by the receiver in the ERP system as "completed". This is a second information transaction, this time as an output by the node. It is classified as: Order: endogenous; Family: cost; Genus: prime; Species: information.
- The ERP system triggers an automatic payment action to be processed by the accounting department. This ERP system becomes a second NODE, of the software type, denoted S and numbered 2S01. The trigger transaction is classified as: Order: endogenous; Family: cost; Genus: derived; Species: knowledge (rather than information because this trigger conveys the understanding the purchase order is satisfied and complete, and the financial impact of the order can now be accounted fully).

■ Now, assume that the shipped materiel is a highly toxic substance subject to regulatory control for its movements and custody status. Assume furthermore that the shipping truck is equipped with a GPS-based tracking system, and that the substance must be tested for turbidity[*] after each transport. The truck's arrival at the plant is automatically signaled by the GPS system one hour prior to arrival. This signal arrives at a receiver at the plant, which becomes a type C node (for control), which we will number 3C01. The alert transaction is classified as: Order: exogenous; Family: cost; Genus: adjunct; Species: datum. Upon arrival, an operator (say 1H02) enters the custody transfer status from the truck to the plant, via an online application (say 4S01), which triggers an automatic alert to the regulatory body about the custody transaction. This trigger transaction is classified as: Order: exogenous; Family: state; Genus: derived; Species: information. Finally, a robot (5M01) performs the turbidity test on the substance—the node type being M for machine. The results of the test are automatically transmitted by the robot to a different module of the ERP system, identified as node 2S02. This upload transaction is classified as: Order: endogenous; Family: cost; Genus: derived; Species: knowledge.

The reader may have noticed the absence of an "intelligence" species in these examples. The latter arises when additional analyses are performed on one or more datasets, to infer trends, patterns, and correlations in the aggregate. The study could be as simple as a statistical analysis of the dataset (for example, turbidity results as a function of container size or shipper names). It may dig deeper into the datasets with more robust data mining applications to determine all variables that exhibit a strong correlation with the turbidity results (distance driven, time of arrival, day of the month, period of the year, supply chain bottlenecks, etc.). The study can go further still, through the use of artificial intelligence algorithms to predict the future variability of the results one month, six months, or one year hence, taking into account broader variables related, for example, to political instabilities in regions where key ingredients of the substance are harvested.

---

[*] Wikipedia defines turbidity as the cloudiness or haziness of a fluid caused by large numbers of individual particles that are generally invisible to the naked eye, similar to smoke in the air. For example, the measurement of turbidity is a key test of water quality. Source: Wikipedia. Turbidity. Available at: https://en.wikipedia.org/wiki/Turbidity. Last modified 18th February 2020.

## The Third Milestone

The baseline will paint a picture that is likely going to be complicated, filled with surprises and holes. The surprises will surely include:

- The magnitude of the volume of information pieces generated by people every day, for which little or no control is exerted, and which are either invisible or unmanaged, or both, from a decision-making perspective.
- The unfathomed extreme lack of structure in the storage and access of these pieces.
- The unbridled proliferation of multiple versions of these pieces floating about at any instant.
- The inefficiency, un-precision, duplicity, and incompleteness of information transacted via static spreadsheets and text documents.*
- The utter lack of real-time datum streams that should be derived from operations.
- The primitive capabilities of the outward-looking applications to capture interactions in real time (web site, customer help lines, vendor submissions, warehousing tracking, transport and logistics progress metrics, etc.).
- The utter lack of software capabilities to analyze these interactions and infer optimization opportunities.
- The primitive level of shop floor automation and/or capability to capture real-time data from their activities.
- The anachronistic features of the information technology framework.
- The absence of advanced algorithms to assist decision-making.
- The mismatch between the information that flows upward to decision-makers and the knowledge needs of those decision-makers.
- The widespread reluctance to even consider powerful software tools for integrity assurance (blockchain), decision-making (neural networks), process optimization (automation and robotics), opportunity and threat trending (machine learning), and new revenue streams (big data analysis).

---

* Any document content that is created and updated by hand is static, i.e., lacking the ability to tap into changing information automatically. Spreadsheets, reports, and other contents generated by desktop applications fall into this category. The same contents become dynamic when the transaction platform is a database (for example). ERP systems are another example of dynamic contents.

The list, unsurprisingly, goes on and on according to the extents of the organization's status quo, which could obstruct even the simplest of changes, such as a new directory structure of a department. Not everything on that list needs to be addressed; indeed, some should be delayed to a later future when the step from now to then is too great a leap for the organization to make. But a lot of items will be indicative of the need to change immediately, if only because their analog opportunity costs are hurting the organization's effectiveness. The question, then, is to figure what to do about what and when (the where, who, and how are assigned later, as discussed in the next chapter). In other words, one needs to *prioritize* the action items into a Priority List (the third milestone). The mechanics of this question are the same for all, *and are applied without regards to the current limitations and constraints of the baseline*:

1. Identify the MUST BE items and place them in the top priority list, as well as the required activation date for each one. Within that group, the exogenous items must come first. These become the action items that are mandatory for the organization.
2. For each item on that list, determine the minimum functional requirements, using the unit transformation process mechanism described in Chapter 3 to quantify the inputs, outputs, and constraints (see Figure 3.4), as well as the *probate* requirements (see *directrix* in Chapter 3 and Figure 3.5). This exercise sets out "what it must be".
3. For each *functional requirement*, define the *functional specifications* of the UTP (attributes and characteristics in Figure 3.4), as well as the sources (which supply the inputs, constraints, attributes) and destinations (which will receive the outputs, characteristics, metrics). The exercise nails down "what it can be".
4. For each functional requirement, determine which existing enablers can be retained and which ones must be changed, upgraded, or superseded by new ones. Here, we have the "what should be".
5. For each item, perform "growth analysis". The "growth analysis" is meant to assess how to transform the "what is" condition of the item into a "what could be". Here is the opportunity to gauge what additional valunomy can be added to the item by judicious selections of advanced digital solutions offered in the marketplace. The analysis will typically uncover more than one potential solution for the improvement, each with its own set of requirements impacting the organization's digital construct. The analysis must be restricted to *technically proven* solutions that are available now and bank account benign. Pie-in-the-sky

concepts and bleeding edge technologies have no place in our digital management transformation strategy (because, as the reader will recall, the transformation must not break the bank).

This mechanic is applied to each item on the priority list *individually* and *independently*. There must be no attempt to combine two or more items at this time in the hope of achieving economies of scales or integrated requirements and specifications. Such an evaluation is integral to the digital management transformation strategy which takes places later, when the *digital construct* is designed (in Chapter 5).

Once the mandatory priority list is fully covered, repeat steps 2 through 5 for the second priority list, which will be constituted with the "what should be" items. Repeat this cycle one more time for the third priority list, comprising the "what is not" items. It is possible to tackle the three lists in parallel, of course, provided that the organization has the heft and wherewithal to sustain such an effort—which is unlikely to be the case if the organization's core business is not rooted in information technology. Three concomitant lists mean three times the staffing level of the *transformation team*, and three orders of magnitude of difficulty in maintaining consistency and convergence across the team's functions over time.

The organization has many options to prioritize the work, if the means or budgets are not available. For instance, step 5 (growth analysis) can be skipped for the time being, which confers the benefit of shorter design times for the digital construct (studied in Chapter 5), but at the price of potentially irrecoverable sunk hardware and/or software costs when future compatibility is obstructed. The scope could be limited to the mandatory priority list (which is really not optional, given the compulsory aspects of these items). The scale of the work can also be dialed back to a single department within the organization (an approach discussed in Chapter 5), which will give the organization the chance to get its feet wet within a constrained, risk-corralled execution. There are, evidently, any number of ways to divvy up the work in a manner commensurate with the organization's limitations. The singular option that should not be kept under consideration is the status quo: doing nothing is not a strategy, it is dereliction.

## *Getting Ready for Part II*

The baseline work addressed in this chapter culminates in the first, complete picture of the capabilities and limitations of the organization from the perspective of its information ecosystem. The picture is by no means complete, since the multiple information components uncovered to date have not yet been mapped out into a single overarching *digital network* (cf. Chapter 2), nor will any be mapped. Indeed, little benefit would ensue from mapping out how things how are now, since the picture will be erased and re-sketched into an entirely different configuration by the time the *digital construct* is designed. Nevertheless, the picture will be compelling enough to the forward-looking manager to envision a promising future, even at this early stage. This motivation to foresee what could be is inherently positive; but it carries with it the danger of jumping the gun to chase after whizbang solutions promising coruscating future returns. Restraint, therefore, is primordial at this moment, lest the need to see blinds the sight. The next step on the transformation journey is backward, rather than forward. One must pause to absorb what has been revealed, in order to succeed in figuring out exactly what can be done with this newfound knowledge. This pause is essential as a starting point for the second part of the transformation by which the sinews of a new business (or operating) model are forged. The exercise requires of management a delicate balancing act between knowing what the status is today but ignoring its limitations, in order to fathom what can be *conceptually* possible when those limitations are obliterated. This will be the essence of the next chapter, where the reader will be asked to hold in mind two opposing viewpoints at the same time.

Luckily for the reader, such a thing is nowhere near as difficult as it sounds.

## Bibliography

Keays, Steven J. *Investment-Centric Project Management: Advanced Strategies for Developing and Executing Successful Capital Projects.* J. Ross Publishing, Plantation, FL, 2017, 419 pages.

## Chapter 5

# The Business Model

*Ask not what you can do with data; ask what data you need to make money.*

Part I of the binary strategy was about what you have. Part II defines what you want, through a business model designed to make your information assets make you money. All is possible in the digital realm; hence, realistic focus is the marching order, guided by the findings of Part I. Part II is framed around revenues, costs, and profitability. It ignores the "how"—a question that is entirely technological, which is the task of Part III.

## What a Business Model Is

### Show You the Money

The reader has an embarrassment of riches when it comes to the literature on starting, running, and growing a business. The management orthodoxy is essentially settled, buttressed by centuries (really) of good and bad precedents. Management principles come and go according to faddish concepts, but the fundamentals remain surprisingly consistent at their core. Consistency imbues them with a universal character that remains equally pertinent to commerce, non-profit organizations, academia, and government structures. For our binary purposes, four of these fundamental principles suffice to frame the discussion.

First principle: All is cost.

This is the self-evident truth that we hold dear, postulated in a cheeky echo to the ancient Pythagoreans.* It encapsulates the fact that, as soon as people join together to pursue a common goal, their actions and inactions will incur costs, be they in time, in labor (free or paid), or in expenses (money outflow). This goes for businesses and governments (obviously), for non-profit operations, and for charitable/spiritual endeavor as well (even a prayer requires time). Nothing more needs to be said.

Second principle: Revenues are life.

This second truth is also a testament to Captain Obvious. Costs flow outward. Something must flow inward to balance things out, and achieve, in the absolute minimal case, zero net flow balance. The life of an organization will come to an end when inflows do not cover outflows. The organization needs *revenues* to offset the cost outflows.

Third principle: Costs can make money.

We will differentiate between two kinds of costs: the *expenditure* and the *expense*. The former is incurred within the *production schema*—defined as the sum of all activities that go into creating that which will generate revenues. Expenditures are tools of monetization, necessary and unavoidable. Examples include production labor, machine time, materials, design, and shipping, to name a few. They can (but may not) appear on an invoice as itemized charges. *Expenditures* are covered by the revenues from the sale. *Expenses*, on the other hand, do not enter directly into the production schema, nor appear on an invoice as an itemized charge. Functions like management, reception, business card printing, marketing campaigns, and the digital asset management system are common examples. *Expenses* are covered by the *profits* resulting from the sale. In this manner, one can view *expenditures* as a measure of the efficiency of the production schema, while *expenses* are a measure of the organization's profitability. One strives to optimize and rationalize *expenditures* but cut and eliminate *expenses*.

Fourth principle: Management is living.

---

* The Pythagoreans were a secretive sect in ancient Greece, circa 6th century BCA, steeped into mathematical mysticism. The sect's famous dictum was *all is number.*

*Costs* are to an organization what entropy is to nature: a relentless proclivity to increase. Revenues on the other hand happen only through concerted effort. Revenues require an explicit *intention* to come into being, whence comes functional structures arise. Profits, recurring and sustainable, require stalwart management. The life of an organization is sustained as a function of management's wherewithal.

## Vision and Mission Are Not Business Models

Business models explain how organizations marshal their resources toward revenue generation and sustained profitability. All business models start with the question "what to sell" and proceeds to design the functional blocks needed to produce the object of the sale (product or service). The model must also answer the questions "sell at what price" and "sell at what profit margin". Getting to the *right* answers becomes the organization's secret sauce. In many management circles, things like vision, mission, strategy, tactics, and value proposition will be included in the business model discussion. But the reader is warned against teetering in that direction. Getting lost in the romance of the vision thing, or the aspirational pursuit of a suitable mission, is all too easy against a backdrop of anticipated greatness enabled by digital powers. Managers and leaders alike should heed these words:*

Never fall prey to romanticizing the nobility of your vision.

That is not to diminish the roles that visions, missions, strategies, and value propositions play in the underlying motivations of an organization. They matter. But they must always be subjugated to the fundamental four principles enunciated above. No inspirational vision can amount to a hill of beans if costs outstrip revenues. When stripped of euphoric proclivities, the vision comes down to a simple statement: "stay in business and thrive"; the mission, "succeed in selling what you offer"; the strategy, "sell at a profit"; and the value proposition, "buyers succeed with what we sell".

The reader may wonder why we should be harping on these matters in this way. After all, the text takes it for granted that the reader understands

---

* See *Investment-Centric Innovation Project* Management, Chapter 8, under the heading "The Vision Thing".

the revenue-cost imperative innately, preaching to the choir as they say. The reason for the insistence stems from experience: when a leadership team comes to the realization that new revenue and profit sources are in the offing from adopting potent digital solutions, it is susceptible to focus its attention on these solutions as the end unto itself, rather than being only the means to the end—which is the conduct of business affairs toward profitability or shareholder returns, or both.

The remedy against this vision-thingy infection has a name: the digital business model, which is defined as:

> The set of tactics deployed to extract the revenue potential of one's information assets, while simultaneously mitigating the costs incurred from the extraction's mechanics and mechanisms.

The reader should note the binary nature of this statement, which emphasizes revenue generation in equal measure with cost mitigation. Formulating the elements of execution that go into achieving both is the business of the digital business model. The model is developed by addressing each of the four principles in that order. The first principle (all is cost) has already been addressed, incidentally, in the previous chapter. All that was discovered in Part I of the binary strategy falls under the heading "costs". The second principle (revenues are life) belongs to Part II of the strategy, i.e., this chapter. In the business model, cost mitigation will figure more prominently than revenues for a simple reason: the number of cost sources vastly outnumber the quantity of revenue channels. The information asset costs also have a much bigger impact to the bottom line over their *economic lifecycles,** because almost everything that an organization does is a cost driver. The reader should note that the discussion must be limited precisely to the aspects of the model that deal with the information assets. The broader issues related to the global corporate strategy, under which falls the digital business model, are not part of our discussion scope: a simple categorization between acquisition costs and operating costs.

---

* The expression *economic lifecycle* is interpreted in this book as the chronological time between the decision to acquire and the effective retirement (de-commissioning, replacement, elimination, destruction). The lifecycle begins when costs are first incurred (associated with the decision to buy) and the time when costs or expenses are no longer permanently incurred.

## To Bit or Not to Bit

Information assets are managed according to usage, which boils down to two types: human and machine. The *human* usage, called the called the *plik\* set*, encompasses the information sources (data, documents, tables, files, folders, forms, printed sheets, etc.) that are either created or used principally by people. Everything that is touched by people as they go about their business within the organization belongs to the *plik set*. *Plik* assets can be printed or digital, filed in drawers or stored in computers. The nature of their medium makes no difference. What matters is that the *plik* exists as an extension of people's organizational functions.

Everything else belongs to the *numer† set*. The *numer set* comprises all components of the knowledge spectrum (Figure 3.2) *whose existence requires electricity*. This caveat is what differentiates human and non-human datasets. For instance, anything on paper belongs to the *plik* set, while everything digital belongs to the *numer* set. Since humans have, in principle, the ability to access the entirety of the organization's information assets, the *numer* set is itself a subset of the *plik* set, as shown in Figure 5.1. This topology highlights the business imperative of keeping humans in the role of ring masters, no matter the sophistication of the *numer set*. The distinction between the two sets is essential because the business model and the management framework differ for each one.

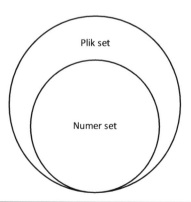

**Figure 5.1    The information asset universe.**

---

\* "Plik" is the Polish word for "file".

† "Numer" is the Polish word for "number", understood in our context to imply "numeric" in the computer sense. All computer data exist as numerical binary representations sequenced into a series of zeros and ones.

# Defining the "Want"

## *Reading the Cards*

Information assets are windows into the organization's holobiome relationships that exist on the strength of their information transactions. Datasets have a story to tell, written in blunt words bereft of biases and motives. These stories can be probed, dissected, and deconstructed once the right language is applied to them. Out of this language comes understanding and intelligence. This, in a nutshell, sums up the triple role of the digital business model: first, to choose what is to be read, for what reasons, and at what frequency; second, to interpret the stories into actionable lessons; and third, to change and marshal the narrative toward a conclusion ordained by management.

Evidently, the stories will be as varied as the organizations weaving them. Stories will go according to script most of the time, with minor bifurcations here and there. At other times, the script will veer off on a tangent, or go rogue of its own volition, notwithstanding the wails of management. Sometimes the script will be re-written by outsiders, with utter disregard for the whims of the organizations. In virtually all cases, and this is the crux of the point here, the impending changes to the script will be already be written in the data *before* they are manifested. The hidden or unsuspected presence of those hints is the reason why a digital business model is essential to the longevity of an organization.

> The digital business model exists to read and mold the story told by the organization's data.

Predictably, each digital business model will be unique. There is no generic template to follow, no 12 steps to this aim or that, no predefined checklist from which to pick and choose seemingly good options. Furthermore, the potency of a business model is *independent* of the technological solutions chosen to do the reading and the interpreting. This caveat bears emphasizing, given the propensity of newcomers to the digital transformation game to becoming spellbound by the glitter of "newer". The model is never about advanced software, artificial intelligence, whizbang applications, or the Internet of Things.

> The digital business model is all about the questions asked of data, and none about the technological wizardry deployed to ask the questions.

In this manner, the methodology for developing the digital business model of your organization is predicated on exploring the *kinds* of questions that management must consider. Each answer will correspond to a "want" instance for the digital business model—which is to be called henceforth as a *model call.*

> The model call is a question re-formulated as an action around its answer.

For example, one may ask what is the volume of orders received each week from a specific regional market. The *model call* rewords the question as follows: develop the capability to quantify the weekly volume of orders for each market served by the firm. Note also the natural extension of the *model call* to a second *call*: develop the capability to quantify the weekly volume of orders for each market not currently served by the firm. The flavor of the questions highlighted below should suffice to give the reader a sense of the answers sought; it is simply not possible, nor practical, to offer a complete list. The lines of enquiries follow the classification system in the following fashion:

- The class of information will be anchored by the highest level, the H class, which corresponds to the organization's senior management.
- There will be two lines of enquiries according the Order of the information. The exogenous order will be considered first, with a focus on revenue-driven questions (revenue family). The endogenous order will be discussed next, with a focus on cost-driven questions (state and cost families).
- *Expenditure* costs will be addressed by questions pertinent to the prime and derivative genera.
- Expense costs will be in turn relate to the derivative and adjunct genera.
- No questions will be asked of the species of information. That aspect is dealt in Chapter 6 (including the associated node structure from Figure 3.3).
- Finally, a set of questions will be raised in the context of the knowledge spectrum (Figure 3.2) to guide the reader on deciding which spectrum index to assign to what *model call.*

Let us look at an example of the process of defining a model call. We recall our fictitious company IFL from Chapter 4. In recent months, most of the requests for proposals (RFP) from various South American clients have sported a change in the type of oil filter mandated. A few of these RFP have also included a change in the electrical design code (from IEEE to IEC). Both changes have immediate ramifications upon the engineering, procurement, and certification procedures of IFL. IFL has yet to undergo its digital management transformation. Therefore, it has no way of clueing into the fact that these two changes may actually be harbingers of two separate trends in specific markets served by IFL. Ideally, the firm would be able to quickly identify these changes as trends, and react to them proactively, using anticipatory management to sign up new suppliers for the oil filter, negotiate preferential pricing agreements, quantify the engineering and certification impacts of the new code, line up the requisite expertise with the pertinent jurisdictional authority, and adapt the project schedules in accordance with the delivery cycles of the new filter. Here, we have at least three potential model calls for the digital business model of IFL:

- The firm's quotation group must be able to detect, in real time, variations originating in client RFP documentation (relative to the same client's past history of orders with IFL).
- The detection must be followed by a real time trending analysis to determine if the variations amount to a trend, which must trigger an anticipatory management response.
- The trend must be analyzed for root causes (why the variations suddenly appeared), and their impact upon IFL's production, operation, and field support schemes.

Note how the three model calls are formulated around the questions "has there been a change in the RPF requirements" and "is the change going to impact the business". Note also that the calls are silent on what mechanics and mechanisms should be used to perform the change detection, the trend confirmation, and the impact analysis. The model calls limit themselves to stating the need to detect, confirm, and analyze changes coming in from the firm's holobiome. These calls could be supplemented with derivative calls as well, by specifying specific actions to take. For instance, one could develop a series of specific *action scenarios* to be carried out in a specific sequence. One of these scenarios could address the procedure for signing up new vendors in different parts of the world. The model call becomes more complex but, assuming a well-structured set of action scenarios, it enables a firm to manage its follow-up actions more efficiently. The model call goes something like this:

Detect => Confirm => Analyze => Enact scenarios x, y, and z.

Evidently, the digital business model is simpler without the scenarios; it lends itself to being compiled as a straightforward list of model calls. The addition of the *action scenarios* involves a greater investment in upfront design; one must spend time and effort to predict the *scenarios* that are likely to occur often for a given call, perhaps leaving out what happens only rarely if ever. Each scenario must be defined in terms of the inputs (yielded by the model call analyses) required to carry out the follow-up action. These inputs, in turn, will speed up the selection of the appropriate digital solutions for each model call. That is the upside of including scenarios in the model calls. The downside? There may be too many *action scenarios* to contend with, yielding a degree of complexity to the overall digital business model that is unmanageable.

## *The Exogenous Model Calls*

The exogenous dataset identifies with the organization's holobiome (customers, suppliers, regulators, stakeholders). Customer and supplier relationships are characterized predominantly by incoming data conveying the revenue story. Regulator and stakeholder relationships generate mainly outgoing data about the organization's state of affairs. The first pair's interactions are highly dynamic and continuously changing, echoing the cash flow flux of the firm. The second pair's contributions are notably slower, passive, and quasi-static, rarely harbingers of unexpected twists and turns. Consequently, the incoming dataset has a disproportionally greater impact upon the day-to-day health of the organization. It should command the lion's share of attention in developing the business model.

The incoming data yield a unique window into your relationship with clients and customers. What stories do these clients tell you? Deciphering them is readily accomplished through the so-called "8Qs", introduced in Chapter 2 of *Investment-Centric Innovation Project Management*. The acronym stands for "eight questions", meant to be asked in the following order:

1. What does the buyer currently buy [from you]?
2. Why?

3. What does the buyer does not currently buy [from you]?
4. Why?
5. What does the buyer wish he could buy [from anyone]?
6. Why?
7. What does the buyer not want to buy [from anyone]?
8. Why?

The questions were originally formulated to help innovators and inventors gauge the commercial viability of their ideas. While their focus was primarily meant to highlight the differences between the status quo and possible novelties, in our context that focus becomes an enquiry into the reasons why your buyer (client, customer, etc.) buys from you instead of someone else today, and how these answers may change in the future. That is the essence of the story conveyed by the incoming data. The answers to each question may not be readily visible to the enquirer (unless you happen to have a monopoly or monopsony on your market). Indeed, those answers will lie hidden beneath data's surface. Uncovering them requires a four-step process executed sequentially: statistical analysis, trend analysis, *opportunity assessment*, and compliance verification.

**Statistical analysis**. The statistical analysis paints your revenue profile and its territory. The statistics quantify what sells or not; who buys what when where; who does not buy from you, wherever they may be. Statistics inform you on the price sensitivities of your various markets, from which you can potential enact on-the-fly pricing adjustments as a function of external conditions (a competitor offering a deal, for instance). Demand volumes likewise inform you on possible just-in-time production output adjustments and product variations. Warranty claims and help line calls can point to sudden or unsuspected issues with the product or with the usage of the product by the buyers. These findings can in turn mobilize your after-sale services group in anticipation of the problems uncovered. They can also serve as triggers to warn your broader client base and head off a public relation crisis, or a product recall before the issue is taken upon by the media. User preferences is another statistical group that can be inferred from online interactions. Information requests and user suggestions can point to the existence of new potential revenue streams based on the data associated with them. The statistics can also outline the nature of the interfaces between you and your clients (how many interactions occur online, by phone, by email, by mail?). Finally, the statistics will inevitably circle back to the performance of your supply chain in ways that may not be visible to you when each vendor is looked at individually.

The *model calls* in this realm operate at two levels, core and insight:

■ At the core level, the model calls will define what statistics should be compiled on what sources, at what frequencies, and to what degree of integration (of the potential variables that directly or indirectly impact the statistics measured).

■ At the insight level, the model calls will define how the core statistics are to be aggregated into meta-datasets, from which effects/impacts are to be monitored (at what frequency) in specified areas such as production, feature product preferences and rejections, market idiosyncrasies, and supply chains to name but a few.

**Trend analysis**. The trend analysis starts with the statistics and searches for evidence of patterns. These can be inferred directly from the statistical metrics (trending curves, standard deviations, correlations, etc.) or from more robust data mining techniques. In the context of the knowledge spectrum from Figure 3.2, patterns are instances of *knowledge*, which influence the kinds of questions posed by the *model calls*:

■ Why are sale metrics different from one market to another?
■ Are the price points heading south or are they amenable to increases?
■ Are sale numbers revealing new buyer whims?
■ Are buyers abandoning you for competing offers? Why?
■ Are there new regional preferences or circumstances that are affecting sales in specific markets?
■ Is it possible to infer customer satisfaction (or displeasure) from available data?
■ Why are products failing at different rates according to different markets?
■ Are marketing materials (and translations) effective in all markets? If not, why?
■ Are there innovations appearing in specific markets that could present a threat?
■ Are customers changing their preferences in the way they wish to interact with you?

These examples are representative of direct pattern inferences; that is, there is a direct link between the data and the deduction. Robust data mining also enables analysts to make indirect inferences across datasets that are not directly connected. For instance, a regional market may be experiencing a switch in customers' product preferences because of supply constraints caused by natural disasters or political wrangling in a producer's country. These patterns are more difficult to confirm but, when ascertained, indicate new opportunities for sales or risk containment. The analyst must exert constant vigilance about the truth of a detected pattern. In medical circles, the expression "false positive" describes the situation, whereby an infection test for example will come back positive, while the patient is actually not infected. False patterns can lure an organization down rabbit holes, where precious time, resources, and capital are flushed away. Hence, every trend analysis must include an independent vetting process for presumed patterns to make certain that the pattern is indeed real *and sustained* rather than representing a transient blip in customers' habits. Inferred patterns are also fertile grounds for confirmation biases emanating from the analyst's internal volitions.* Confirming a pattern will involve running a separate trend analysis on the same dataset, and perhaps complement the hunch with market research on the ground. When a pattern is confirmed (be it negative or positive), it is followed up with an *opportunity assessment.*

**Opportunity assessment.** The opportunity assessment is to trend analysis as knowledge is to intelligence on the knowledge spectrum. The confirmed pattern is the starting point. From there, one seeks to map out the execution strategy to address the ramifications of the pattern. Even if a pattern is negative (say, a product offering is losing market share), the opportunity lies in either doing something to stop the losses, switch product offerings immediately, or abandon the market altogether. In all three cases, the opportunity will save the organization money. If the pattern is positive, the opportunity represents increased sales, higher profits, market expansion, or other options to improve the bottom line. The opportunity almost always impacts one's supply chain and the logistics that surround it. And in all cases, the assessment creates the conditions for the organization to manage itself by anticipation, rather than react (usually too late) to changing landscapes. Finally, the opportunity assessment opens up management to the possibility of interacting more directly with its holobiome, especially in the realm of commercial partnerships and joint ventures.

**Compliance verification**. Compliance verification targets the "state family" transactions of the organization. As we noted earlier, the nature of

---

* Confirmation bias is discussed at length in Chapter 10 of *Investment-Centric Innovation Project Management.*

the "state" datasets is sloth-like. Dealing with regulatory bodies, with shareholders, or with government bodies, is characterized by inertia—processes, procedures, and interactions are slow, ponderous, immutable, rigid. This catatonic environment frames the pertinence of the *model calls* in terms of punctilious adherence to formats and data structures. Irritatingly, the transaction framework is usually inefficiency (if not outright retarded), ineffective, and technologically archaic. Thus, these transactions are inherently wasteful, expensive, and immune to capital market pressures. In order words, they are profit wasting for the organization. One achieves nothing by inveighing against the technological ossification of these bureaucracies. The only control available to the organization is with the preparation works that precede such transactions (such as tax filing, regulatory reporting, quarterly earnings calls, etc.). That is where the *model calls* must be focused with such questions as:

■ For recurring filing requirements, what contents can be assembled/created automatically, rather than by manual labor?

■ What contents can be created once and re-used at will?

■ What contents are currently created by hand (spreadsheets and text editors) but could be automated?

■ When online/electronic submissions are available, how can the upload process be automated, rather than require manual handling?

■ What submissions could be done without printing?

■ How can mandatory reporting events (requiring disclosures to the pertinent authorities) be detected before they occur?

■ How can follow-up to action items dictated by previous filings be automatically tracked and acted upon without manual interventions?

■ Can the organization's submittal process be digitalized to overcome the inherent deficiencies of the governing jurisdictions?

The reader may have noticed a common thread woven through these questions: the need to get away from manual, labor-intensive handling by humans, which are the primary expense drivers to the organization. Despite the infuriating antiquated mechanics of regulators and governments, aggravated by their ubiquitous reliance on the printed document, filers and reporters can still inject production efficiencies into their "state"-related activities. The *model calls* must be expressed in terms of finding and achieving these efficiencies. You will make no money, but at least you will cut your expense costs and improve your bottom line.

## Endogenous Model Calls

The endogenous datasets speak to the organization's cost structure, whether they are expenditures or expenses. If the former's case, one seeks efficiencies, optimizations, and throughput improvements. In the latter's case, one seeks to reduce, mitigate, and, ideally, eliminate them outright. Whenever a human is involved, the risk of re-inventing the wheel is omnipresent (creating a standard content letter from scratch on a laptop, rather than issuing it from an automated document generator). The risk of datum hoarding is equally pervasive (through individual spreadsheets rather than operating from a common database application). Consequently, the *model calls* for the *plik set* are geared primarily toward *expenses*—i.e., the elimination of manual labor associated with contents creation, data aggregation, document distribution, version control, information waste, and procedural warts. Typical questions raised in this context will include:

■ What content creation processes can be automated? Digitalized? Streamlined?
■ Which information sets can be moved from static to dynamic?
■ How are dynamic contents compiled, tracked, and reported? Which should be digitalized (moved from spreadsheet to database applications)?
■ What filing processes can be moved from static to dynamic uploads?
■ What templates and standards should be developed?
■ How is version control, for what *plik* component, performed? How is access control executed?
■ What can be migrated from print to electronic display?
■ How can signatures be migrated from print to numeric?
■ How can conference rooms be digitalized such that participants annotate presentation materials digitally instead of written emendations?
■ Which submittal/reporting processes can be digitalized? Automated? (Things like internal management reports, commercial bid invitation packages, bid submissions and comparisons, invoice submittals, third-party deliverable uploads, cost estimating and capturing)
■ What production, operation, and warehousing records can be digitalized (data, testing, calibration, licenses and permits, import/export records, logistic mobile data)?
■ What paper forms and clipboard checklists can be eliminated and integrated online?

- How are materiel/certificates/personnel qualifications tracked and managed for traceability and audits?
- On the production floor, how to eliminate paper completely, including QA/QC records, drawings and technical documents, parts tracking, measurements and calibration, change management, personnel movement tracking.
- How can printed processes, procedures, policies, forms, and instruction sets be migrated to a digital delivery system without any paper remaining?

The *model calls* for the *numer* set target *expenditures*, with a view toward monitoring real-time throughput metrics, inefficiency drifts (a machine getting out of calibration for instance), production bottlenecks, utility consumptions (electricity, power, consumables, etc.), and performance degradations (product quality, reliability losses). The *numer model calls* differ from their human kindred in one additional way: they operate at three functional levels,* compared to one for the *plik set*. First is the *system* level, where individual machines, sensors, and other non-human data generators are found. Systems act as unit transformation processes of data and records (pursuant to Figure 3.2) into line operation outputs. At this level, the *model calls* target the inputs and outputs of these UTPs, both extant and non-extant. The types of questions involved with systems include:

- Is the system handling the input and output data streams acting passively (receive and send data only) or actively (additional data processing performed on the inputs and on the outputs)?
- Is the system capable of communicating its functional metrics (the role(s) for which it is used) and the exceedances to its design limits? If yes, is it equipped to do so? If not, should it? If not, where is this information captured and processed?

---

* This three-tier categorization is borrowed from the concept of *feature space* developed in Chapter 4 of *Investment-Centric Project Management*. The concept starts with the physical configuration of an asset (a plant, a facility, an airport, a shopping center, etc.) as the underlying foundation for managing capital projects. The configuration places the *plant* as the highest domain of the project (the thing that will generate revenues for its shareholders). The *plant* is made up of two or more *installations* and with each installation performing a set of related physical functions. Each *installation* is in turn divided into two or more *systems*, integrated to work together seamlessly, where one system performs one or a few specific functions. Each system is broken down into two or more components, which are the constitutive bits and pieces—the screws, the power cord, the base, the dials, etc.

- Is the system capable of communicating its state (on/off, operating settings, reliability metrics, calibration/alignment drift, future maintenance needs, etc.)? If yes, is it equipped to do so? If not, should it? If not, where is this information captured and processed?
- Is the system capable of communicating utility information (power draws, consumable consumption, noise generated, emissions and effluents, greenhouse gas [GHG] contributions, etc.)? If yes, is it equipped to do so? If not, should it? If not, where is this information captured and processed?
- Is there paperwork attached to the system that must be regularly updated (calibration checks, quality control checks, maintenance records, part replacement registers, etc.)? If yes, can the paperwork be digitalized? If not, why? If not, where is this paperwork kept, by whom?
- Are there manuals, data books, parts lists, troubleshooting guides, drawings, and other similar information attached to the system? Must they be updated, at what frequency, by whom? What are their sources? Can they be digitalized? If not, why? If not, where are they kept, by whom?
- Is there a virtual model of the system to simulate its operations, its reliability, and its hazard containment? If yes, who can access it? Where is it located? Is it accessible in real time? Is it integrated into the plant's model? If not, why?
- Is the lifecycle history of the system documented (capital vs operating costs, maintenance activity, reliability record, comparison of metrics between like systems, comparisons between alternate design/brand systems, vendors and suppliers, logistics costs, etc.)? If yes, is it included in the *plik set*? If not, does it exist? If not, why? If yes, can it be digitalized and automated? If no, why?
- What are the overlaps between the system's contributions to the *numer set* and the associated elements of the *plik set*?

The second level of the *model calls* operates at the *installation* level, where related systems are integrated and rolled up. The installation level is where revenue generation begins for the organization. Here, the *model calls* are formulated principally in terms of integration efficiency, throughput bottlenecks, and production volumes expressed in records and knowledge (pursuant to Figure 3.2). All the system-level questions can be posed to the installation as well. Additional questions arise from performing trend analyses and opportunity assessments on the installation as a single entity.

The final level rolls up the installations into the *plant*, where the mind is focused on profitability and compliance. At the *plant* level, one seeks knowledge and intelligence (pursuant to Figure 3.2) over the organization's operations. The *model calls* are prescribed in terms of the knowledge spectrum required to perform the trend and opportunity investigations using real or near-real time information. The *calls* also probe the interactions between humans and machines, and the interfaces between the *plik* and *numer sets*. Vectors of enquiries can include:

- Are human data transactions visible, traceable, and verifiable?
- Are machine data transactions visible (to humans), traceable, controllable, and auditable?
- Are audit processes and procedures digitalized (rather than rely on paper) via database applications?
- Are both *plik* and *numer sets* properly governed in terms of data ownership, access, protection, and recovery?
- Are there mechanics and mechanisms in place to connect the endogenous *model calls* and the exogenous transactions originating in the organization's holobiome?
- Is the digital construct capable of executing all *model calls*?

# Execution Guidelines

## *Setting the Digital Construct for Success*

The *model calls* establish the conceptual framework for exploiting the organization's information assets. They are enunciated without *any* attachment to one technological solution or another. Nevertheless, some technological considerations must be addressed to frame the pillars of the digital construct to be developed in Chapter 7 to turn the conceptual framework into an actionable management structure. These considerations set the digital direction of the organization. Once these pillars are positioned, one can proceed with developing the design details of the entire architecture. Keep in mind that there are no right or wrong decisions in going one way or another; each path represents a set of unique capabilities and limitations, corralled by cost and complexity ramifications. What matter, at this point in time, is to make those decisions *before* you embark on the development of the solutions supporting the *model calls*.

## Silos or Beaches

The expression "datum hoarding" appeared earlier in this chapter to describe the risk carried by people's propensity to rely almost exclusively on desktop applications to create documents, files, and spreadsheets during their daily activities. Convenience and immediacy are obvious reasons for their preponderance. However, this individual benefit comes at the expense of the organization. The *modus operandi* yields information silos, content invisibility, stranded information assets, rampant version proliferation, digital waste, and uncontrolled content modifications.* Even when shared folders are used, the information they contain remains invisible to all but those who created it. The point is this: organizations stand to save enormous operating costs by automating the process of content generation and manipulation. Switch from spreadsheets to databases and from text editors to online document compilers. An example will suffice. In the construction industry, large contracts are typically awarded through a bidding process that begins with the issuance of an RPF or an invitation to tender (ITT). The package will include a scope of work document (customized), the commercial terms (fixed), the technical specifications (fixed), the drawings (customized), the health and safety plan (fixed), and the quality assurance plan (fixed), as well as a number of supplemental procedures and templates that are fixed (fixed = does not change according to the project, customized = specific to the project). The commercial terms will usually be broken down into terms and conditions (fixed), and a series of template tables for bidders to enter their price information. The latter will be issued as blank spreadsheets, each containing multiple worksheets. Bidders return this plethora of completed documents, which will then be assessed by the owner's bid team for compliance and comparison. The cost tables will be assembled into enormous spreadsheets with which the so-called "bid spread" will be done. All of this, of course, will be done manually, via email or worse, mail. Imagine now that the bid package is assembled via a database-enabled web application, with the many tables and bid answers entered online by the bidders via the same web application. The comparison analysis would then be done automatically by the system, all cost details checked at once by the system, and bids scored automatically by the system using a predefined list of criteria. What's more, all of this information would be instantly visible to whomever is authorized to see it within the organization.

---

* The plight of Excel, in particular, is explored at length in Chapter 17 of *Investment-Centric Project Management* (under the heading "It's a spreadsheet world after all").

This is the first decision that the organization must make before embarking on the development phase of the *model calls*. Do you continue to run your organization as a far-flung silo farm or do you bring everyone on the beach for full access to your sea of information?

## In or Out

The second decision to make pertains to the physical location of the organization's information tools. If the first decision sticks with the silo layout, the second decision is irrelevant. But if it embraced the beach, the decision matters. Organizations have traditionally kept their information assets in-house, inside data servers behind closed rooms. The arrangement works because so much of those assets end up on people's computers and filing desks. This architecture is incompatible with the beach approach. The information assets must be, at the very least, stored centrally (in the data server sense). But centrally does not mean one physical server in one room. It means, *a priori*, a database architecture, regardless of where this database resides, which can be local (physical servers on the premises or networked), remote (via the cloud or some other third-party service provider), or distributed (blockchains and other distributed ledger technologies). As a starting position, a combination of the two should be contemplated. Keep in-house the information assets that are strategically critical to the organization (financial data and backups, core knowledge and proprietary knowhow, etc.). Everything else can reside outside of the physical confines of the organization to reduce your IT footprint and operations costs. Keep to an absolute minimum the information that should reside on individual machines.

## Own, Lease, or Subscribe

Another consequence of choosing the beach over the silo is the possibility of eliminating ownership of the hardware and software deployed within the company. There is an embarrassment of riches in the marketplace when it comes to leasing and subscription opportunities. Data storage can be completely outsourced to the cloud or a third-party provider. Software-as-a-Service (SaaS) eliminates the need to buy (and maintain) costly licenses. Processing-as-a-Service (PaaS) does the same thing for high-end expertise (such as artificial intelligence, data visualization, machine learning, distributed ledgers, operational data streaming, etc.), freeing up the organization from the costly expenses associated with starting and maintaining specialized internal groups. Access to large datasets (to train neural networks for example) via leasing or subscription agreements is also becoming a viable business tactic. The list of options and applications is too vast to corral within this space. Suffice to say that it pays to take the time to think about what is worth owning (much less than you would expect) or not.

## Security

No matter where the decisions from the first three questions, the fourth and final decision affects every organization of every type: what to do about data and privacy protection. The moment that the organization is connected to the internet, the issue of digital security enters the fray, and bluntly one might add. Hacking, data theft, asset corruption and destruction, and a host of other nefarious acts by insiders and outsiders are no longer the concern of large corporations. Every organization is likely to suffer cyber attacks at one time or another. It is not a question of "if" but "when". Consequently, all digital transformation decisions must make allowances with the requisite provisions to achieve stalwart data security. Avoiding the public sphere is not a solution either: staying private across all your holobiome transactions will not stay the dangers of digital offensives against you.

## Final Words of Caution

The formulation of the *model calls* is an endeavor best suited to the organization's management team. The size and scope of the operation will bear directly on the number of calls made. The reader is cautioned against running up a list of calls numbering in the hundreds; such a number will dramatically increase the complexity of the digital construct that will be designed over the next two chapters. A good rule of thumb is to aim at seven to ten *model calls* for each principal department/group of the organization—corresponding to the level immediately below the H level in the classification system of Table 3.1. If the list goes much beyond ten, resolve to divide them into three subsets: the "A", "B", and "C" subsets. Retain only the "A" subset for the implementation phase (discussed in Chapter 7). Plan to undertake the "B" subset once the organization has had time to get comfortable with the inner workings of the newly implemented digital management framework (which will take a few months at least). Then, repeat the exercise with the "C" subset later in the future, taking care in the meantime to revalidate the items against the experience gained hitherto.

The plant-level *model calls* for the *numer set* will require attention from the CEO and COO, given their impact over the organization's overall efficacy.

Finally, let us reiterate one last time the importance of conducting the *model call* formulation without prejudicial considerations of the technological solutions that might do the job. A lot of what passes for leading edge, innovative digital solutions will not have reached the maturity necessary to reduce your implementation (read: budget) risks. It will always be possible to find a technology in the marketplace to embody the intent of a given *model call*. Therefore, there is no urgency to tie yourself down to a specific one because it happens to be the "it" thing of the moment in the business literature. Remember that no technology will amount to a hill of beans if you do not understand *conceptually* what it is that you want to get out of the *call*. The old adage "garbage in, garbage out" still reigns supreme in the Symbiocene age.

## Bibliography

Keays, Steven J. *Investment-Centric Project Management: Advanced Strategies for Developing and Executing Successful Capital Projects*. J. Ross Publishing, Plantation, FL, 2017, 419 pages.

Keays, Steven J. *Investment-Centric Innovation Project Management: Winning the New Product Development Game*. J. Ross Publishing, Plantation, FL, 2018. 309 pages.

# Chapter 6

# Binary Tools

*From the perspective of framed painting waiting to be mounted on a wall, a $17 screwdriver is far more appealing than a $13 000 jack hammer.*

Part III of the binary strategy turns what you want (called out in Part II) into what you can use. Picking the right technological solution, it goes without saying, is arrantly important to the health and bottom line of the organization. Which can mean, incidentally, picking nothing new or mere basic, if that is the right answer.

## The General Approach

### *Form Follows Function*

In the previous chapter, the *model calls* were explored and formulated within the context of addressing the matter of what the organization is to become when it grows up from analog teenager to digital adulthood. In this chapter, the discussion segues on toward the question of "how", as in how to turn those *model calls* from concept into actionable reality. At the end of this chapter, the reader will have compiled a list of specific *plik* and *numer* technological solutions (but leaving brand name selection off the table and

in the capable hands of the reader*). This chapter also abides by a caution-
ary guidance akin to Chapter 5's exhortation to exclude technological considerations when formulating a *model call*, namely:

> When choosing a specific technological solution, form must always
> follow function.

This design principle[†] was discussed at length in Chapter 5 of *Investment-
Centric Innovation Project Management*. The greatest—and costliest—error is
to preemptively choose a technology then attempt to fit one's circumstances
to it. Algorithmic sophistication can wreck an organization that does not need
it, an issue that is exacerbated when the capital investment is frightening,
and decision-makers become desperate to prove the merit of the decision.
Sometimes, pen and paper will be best! This is where the guidance comes into
play, as both protector of one's capital and deliverer of valunomic capabilities.

Simply put, you get to the best solution by first understanding what is
really required to achieve the desired outcome—using the *unit transformation process* (Figure 3.4) as the principal decision-making tool. The UTP is
applied in reverse, from *outputs to inputs*, to discover the specific details of
the implementation. In many cases, the solution to a specific *model call* will
be self-evident, thereby bypassing the need for a UTP analysis (for example,
when moving from paper to a desktop application). On the other hand, as
soon as the *model call* implies more than a format conversion, the UTP tool
is advisable.

The UTP is applied as follows. One starts with the *outputs*, which correspond to the *model call*. The *outputs* are further defined in terms of

---

* This book makes no attempt to compare vendor offerings, nor endorses any particular one. In
many cases, the reader will already have a preference (Windows vs IOS vs Java, SAP vs Oracle,
AutoCAD vs Intergraph, Office vs Google Docs, etc.), at which point the issue is about available
capabilities and integration potential. These questions are too far down in the weeds of one's
unique circumstances to be adequately covered in the text.

† "The pilot-tested product lives and dies by the dictum *form follows function*. To its user, the value
proposition is what it [the product or service that was bought] does rather than how it looks. It is
bought first and foremost for its ability to perform its functions as advertised. The feel, look, color,
surface finish and pleasant user interface, to name but a few, are of secondary concern to the
buyer unless these things perform a function as well. The color of a circuit breaker assembly is of
no importance to a buyer, except when that color is tied to an amperage scale, for example. That is
not to say that the presentation and packaging of product is immaterial, however. On the contrary,
no product can afford to create an impression that it is inferior, weak, unsafe or unsuited to its
purpose. *In fine*, the look and feel of a pilot-tested product (the "form" in the above dictum) comes
into play once its purpose is functionally complete".

measurable features by which one can determine the usefulness of the *outputs* (type, format, presentation, contents, etc.). Next, one needs to uncover the *processes* that are required to produce these outputs. From the latter, one extracts or infers the *inputs* needed to perform the transformation, which are defined similarly to the *outputs*. Together, these three sets of definitions yield the solution's *functional requirements.*[*] The requirements must be turned into *functional specifications*, which put numbers and quantities on the *inputs, processes,* and *outputs*. The specifications require the identification of the *constraints* (both endogenous and exogenous) that circumscribe the UTP. The *constraint* specifications can also shed a light on which *attributes, targets, characteristics,* and *metrics* are mandated by the constraints (which will be, for the most part, exogenously driven).

## *Solution Evaluation*

The functional requirements and specifications (together as the *functional set*) form the assessment basis from which technological solutions will be evaluated. Together, the pair correspond to the "function" component of the "form follows function" guidance statement. Through them, an assessor can immediately determine if a given technology or software will be capable of producing the desired outputs and satisfy, by the same token, the underlying model call. It should clear by now why form must follow function: fitting the functional set to the arbitrary nature of a pre-emptively selected solution is arrantly difficult or downright impossible. The outcome will, predictably, be plagued by compromises, shortcomings, and limitations, leading inevitably to frustrations and extraneous operating expenses. This why, fundamentally, form must follow function. It is not the software that matters, it is the nature of the problem to be solved.

> The inversed sequence, whereby a solution precedes the UTP method, never delivers the expected financial benefits.

Complex calls will obviously result in complex *functional sets*, which may foment a situation where no single solution can be found. In that case, the assessor may want to revisit the *functional set* to whittle them down to what is essential vs desirable—keeping in mind that integrating and managing

---

[*] The distinctions between functional requirements, functional specifications, and design specifications are explored at length in Chapter 12 of *Investment-Centric Project Management*.

two or more commercial solutions entail its own level of complexity that drives up *total cost of ownership** (TCO).

The TCO criterion must be an integral component of the assessment process. Technological solutions anchored to a software engine are notoriously more insidious than run-of-the-mill equipment subject to straightforward depreciation rules. Up front capital costs can be either high or low in such a way that they obfuscate their true economic exigencies over their economic lives (measured by their valunomy). The latter comprises the selection process (with its own cost structure), the initial purchase, implementation, personnel training, roll-out, maintenance, annual licenses, back-up and recoveries, upgrades, expansion, platform migration (when hardware becomes obsolescent), partial substitution, obsolescence, abandonment, data conversion to the latest format, and final archiving of remaining information assets. Software is always accompanied by the potential danger of data incompatibility and systemic orphanage (the age-old tension between proprietary and open source *modus operandi*).

> At the enterprise level, the purchase price cannot be allowed to govern the selection decision. Valunomy must be the criterion.

## Post-Solution Processing

The selection of a specific solution signifies the end of the guessing game and the start of the transformation implementation. What comes next is the *design phase*, where the *design specifications* are drawn up to prepare the organization for the solution's actual introduction. NOW IS NOT THE TIME TO BUY THE SOLUTION. It is the time to figure out the ramifications of the solution BEFORE it is bought. The *design specifications* round up the formulation of the outstanding elements of the UTP, namely the *attributes, targets, characteristics, metrics*, and *enablers* (the first of which is, evidently, the solution proper). These specifications establish the physical, control, and algorithmic characteristics of the solutions *as it will appear when implemented into the organization*. The *enablers* will include the people involved, directly and indirectly with the solution; the interfaces with non-human nodes; the operating philosophy, procedures, and processes; and the operating budgets. The *attributes* and *characteristics* comprise the

---

* See *Investment-Centric Project Management*, Chapter 2.

*constraint-derived* functional specifications; the internal variables posited by the organization in relation to how the solution will work; and the utilities (permits, power, physical design features) needed to operate the solution The *targets* and *metrics* speak of the performance of the solution in delivering the *model call(s)*, its operating reliability and maintainability, and its financial encumbrance.

The sum of the functional requirements, functional specifications, and design specifications constitutes the *systemic definition* of the solution. The systemic definition becomes the foundation of the digital construct (discussed further in this chapter) and its primary configuration management tool. The completeness of the *systemic definition* will be gauged by the number of missing pieces encountered when the solution is physically implemented (the subject of the next chapter). It is encapsulated in the schema of Figure 6.1.

Note that the benefits of the *systemic definition* extends beyond the cost efficiency of the implementation phase. They will be prominent when future modifications are considered to keep up with changes to the organization's holobiome. Hence, it is important that whatever changes, additions, or deletions were made during the implementation phase are captured into revisions of the *systemic definition* documentation (including explanations/reasons for the emendations). As a matter of fact, such amendments will

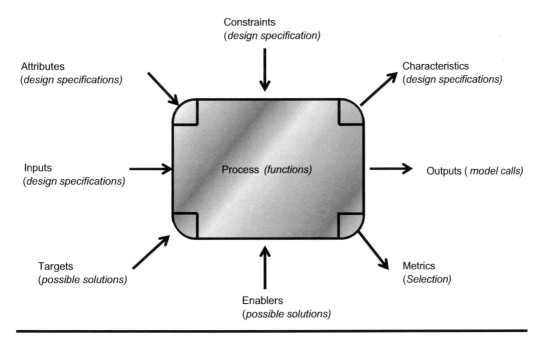

**Figure 6.1 Solution selection. The design principle *form follows functions* is embedded into the unit transformation process.**

necessarily occur throughout time, making it imperative to maintain the *systemic definition* current at all times.


# Plik Solutions

## *The Prime Plik Directive*

The Plik realm is governed by an ideal encapsulated by the following *prime directive*:

> Plik data should be created once, stored uniquely, and transacted referentially. It is the ONLY management process capable of guaranteeing real quality control of plik sets.

Why qualify it as an "ideal"? Because, notwithstanding the correctness of the directive, its implementation is harder to control in real life, where human action dominates. It states that the contents of a datum or record (pursuant to Figure 3.2) should exist only in its current version as a single instance. The caveat implies that other copies should not be in existence. This uniqueness gives it the seminal character of being "relied up" and hence, authoritative and unequivocal in the moment. The character of "being relied up" confers upon the datum/record the attribute of "information asset", in a financial sense. Once created and declared "current", the asset should only be utilized/transacted/transmitted/circulated by reference or linkage back to the source. This is the ONLY way for an organization to GUARANTEE the precedence/authority of the datum. Otherwise, one can never be certain that the version on hand is actually the latest one, a problem that is especially widespread with spreadsheet files shared through emails. Uniqueness, on the other hand, does not prevent copying parts of the source asset into other plik components. For instance, a corporate standard is a unique asset, whereas a particular article of the standard could be copied into a separate instruction manual to explain the implementation of that particular aspect of the standard.*

---

* The principle of uniqueness of existence is tacitly embedded in the creation of dimensional engineering drawings, where a given dimension appears only once on a specific drawing but mentioned elsewhere on this and other drawings through the REF callout. In this manner, if the dimension were to be changed in the future, only the drawing where the source callout appears would be modified. Otherwise, every drawing using the dimension would require a change. Quality assurance in the latter case is nearly impossible.

The prime Plik directive is easy to accept in theory (especially in the context of a stalwart quality assurance framework) but harder to abide in practice. Consider the ubiquitous example of a spreadsheet containing daily production data. The original file will have been compiled by one person (hopefully the same person every time) then shared among various people within and without the department. Sharing will, of course, be via an email attachment, where changes could be made by recipients, who will in turn share their changes dynamically, in some form of informal feedback loop. Once in a while, someone will copy and paste a cell that contains a formula and make a mistake in its expression. That mistake will now propagate throughout the department, via more email transactions. Quality control, in this all-too-familiar example, is IMPOSSIBLE neither in real time or retroactively. Now, assume that the prime directive is in play. The source version of the spreadsheet will be stored, formally, onto the department's network directory. Participants will be invited to consult the source version by email using a link or other referencing mechanisms. Whoever wishes to make changes to the source version will have to either 1) provide the changes by email to the originator or 2) utilize a built-in change management process, germane to the directory's access protocol. In the event that the file must be shared to an external entity (a vendor, a regulator, a lawyer), the same mechanics of referencing will be used; which means that the organization must have a way of controlling exogenous transactions performed on its internal plik management system.

> The prime Plik directive fundamentally changes the ways an organization manages its plik assets. Structurally, datum storage, access, and modifications can only be executed within a database framework, rather than a network directory structure.

The adoption of the prime directive forces the organization to manage its information assets via a formal plik management system (or document management system, in its simpler incarnation). Wanton document creation by anyone at any time will be severely constricted—which may lead to frustrations by staff accustomed to running their own, individual file management systems from their desktops. This may come as a shock to the culture of the place, and usher in extra operating expenses incurred from the exploitation of the plik management system. Notwithstanding, the upside of the directive dwarfs the inconveniences wrought upon the troops: enforceable quality assurance and control; reliable information assets; auditable and traceable asset holdings; and massive operating cost savings over the long run.

From this point forward, the discussion will assume that the prime Plik directive is adopted by the reader. What follows next is a summary of effective plik solutions that are applicable to a wide array of common *model calls* associated with human information assets.

## Low Hanging Fruits

Organizations face a plethora of opportunities to digitalize their information asset processes with minimal capital expenditures and impacts to their organizational structures. These solution opportunities are overwhelmingly weighted toward expense reductions, throughout efficiencies, productivity increases, and enhanced quality management efficacies. The opportunities comprise (but are not limited to) the following suggestions:

■ From paper to computer files. Very few plik assets require a mandatory paper substrate, among them legal documents, mortgages, liens and titles, court depositions, and regulatory filings. For those assets, there is nothing more to be done, save digitalizing the cross-referencing of the physical documents to their scanned counterparts. *For all other plik assets, paper can and should be eliminated completely and replaced with digital documents.*

■ From files to templates. A huge majority of plik assets transacted daily by an organization are similar in contents, format, and presentation (think letters, reports, forms, tables, drawings, technical sheets, contract packages, purchase orders, inspection reports, inventory status, materiel tracking, mileage and expenses, etc., the list goes on and on and on). A simple rule exists for this group of plik assets: if the asset is first created from a previous example, or through copy and paste from a similar exemplar, that asset should be developed into a template, modifiable in contents according to the particulars of the instance. The template becomes the source version, pursuant to the prime Plik directive.

■ From checklists to browsers. An equally significant number of routine plik assets fall under the category "checklist", completed at prescribed recurring intervals by individuals. Quality control, inspection reports, delivery records, mobilization planning, submission verification, and confirmation of attendance/skills/training are typical examples. Unless the checklist must exist on paper, in accordance with a legal or regulatory prescription, it should be converted to an online form (skipping the middle step of converting it into a spreadsheet template), accessible by phone or computer, in real time or in batch mode (when the internet is not available), *and signed off digitally.* Enough said.

■ From reports to browsers. The "plik" category shares the same characteristics of the "checklist" category. These are documents that are generated each time a specific action is taken to monitor progress and/or quality of execution of a prescribed scope of work or to monitor daily, weekly, monthly, or quarterly performance—among many other similar applications. And, just like the "checklist" category, they should be all digitalized via online forms operated the same as those for "checklist" forms.

■ From ink to link. The requirement of a signature (or authentication by stamp and such) is often used to justify paper plik assets. Once again, unless an exogenous mandate prescribes a signature by hand in ink, all signature-bearing documents should be digitalized in such a way that the signature is imparted electronically. Several robust commercial protocols and solutions already exist in the marketplace to solve this hindrance in short order.

■ From spreadsheets to databases. Spreadsheets fall under the "tabular" category. They pullulate across organizations of all sizes and ilk. Get rid of them. The rule of thumb is: if a spreadsheet is regularly updated OR is used to compare datasets OR keeps a historical record, replace it with a database application.

■ From spreadsheets to math sheets. Still under the "tabular" category, except for its usage as a calculation scheme (or template). Here, the issue of intellectual property comes to the fore, since such calculations are typically developed internally by the organization. Not even password protection can protect their contents or stop them from being shared/copied illicitly. The solution? Adopt mathematical software (many exist in the marketplace) to replace them. In addition to enhancing the intellectual property protection, these applications make it possible to fully audit the calculation algorithms and the equations, from which proper quality assurance can be conducted.

■ From blueprints to models. Drawings, blueprints, and technical datasheets are found wherever design, fabrication, and construction activities take place. Drawings are generated directly from a CAD application or from a solid model application, and then, alas, printed for distribution, reviews, and actions. Eliminate the paper format. Insist on transacting drawings electronically. On the fabrication floor or the construction site, skip them entirely and rely instead on the solid model directly.

■ From directories to the plik management system. Everything mentioned above (forms, records, checklists, reports, documents, spreadsheets, drawings, models, etc.) typically exist in directories created on the organization's network. These directories are rendered superfluous when the source data are mapped over to a browser interaction or a database query. The trick is to implement a potent (yet intuitive) record management system to store and access everything. Do away with network directories, department folders, and project-specific directories. Go directly to record management system. Do not Save As. This way, the prime Plik directive can be implemented and enforced uniformly from the get-go, from here until eternity.

■ Beyond email. Transacting plik records via email was already discussed in the earlier example. At issue in this case is to STOP attaching documents to emails. Send links and references back to the intended attachment's location on the Plik management system.

■ Beyond CRM. Customer relationship management (CRM) are applications used widely by sales, marketing, and business development people. They are superb tools for managing exogenous relationships. Why not employ them to do so for endogenous connections as well, and by everyone in the organization? They are far superior to emails for information management.

■ Never print again. All marketing brochures, price sheets, sales materials, user manuals, technical datasheets, business cards, and other materials intended for an external audience are to reside on the organization's website as downloadable files. Never print these information assets again. Furthermore, maintain a live link between a downloaded file and its internet destination so that any update and change made over time to that file is automatically notified back to the address of the first download. This provides the organization the opportunity to touch base with an existing connection of its holobiome, which creates a wonderful marketing or sale opportunity.

## Advanced Plik Solutions

The low hanging fruits can be implemented rapidly, without great expense or organizational re-structuring. They yield the most bang for the starting buck. However, the salient bottom line benefits lie with the advanced plik solutions, which require greater planning and investment. Whereas the fruity solutions will not, for the most part, require new software purchases, the advanced ones will.

> WARNING: DO NOT UNDERTAKE YOUR OWN APPLICATION DEVELOPMENT—unless you are into employing a regiment of costly developers for a year or two... Buy commercial, off-the-shelf solutions; even if consultants are required to customize the solution to your needs, the final tally will be cheaper, and faster, than programming everything yourself.

In addition, the advanced solutions will be deployed across the organization, a move that implies formal management structures, additional IT oversight, bespoke training, and recurring maintenance/license fees. Bottom line: advanced plik solutions must be undertaken as "projects", in the investment-centric project management sense. Finally, note that the fruity solutions were aimed mainly at data and records (in the context of Figure 3.2); the advanced solutions move the focus to the right of the spectrum, towards knowledge and intelligence.

- From templates to document automation. The preponderance of common contents in an organization's plik assets underlies the justification for the prime Plik directive. But templates and reusable contents remain fundamentally static in their users' eyes. The next logical step is to migrate toward a dynamic paradigm, whereby complete documents can be assembled automatically, according to pre-defined presentation schemes, while simultaneously permitting content customization where required. Contract documents and project scope definitions are ideal candidates for this level of automation. Keep in mind as well that these solutions are perfectly suited to an environment where source versions are utilized as the building blocks of a plik management system.*
- From database to dynamic knowledge sets. This one extends the document automation franchise to tabular data, which were converted from spreadsheets to database applications, as low hanging fruits. Data

---

* These applications have progressed in leaps and bounds since the early days of primitive SGML and HTML applications. The reader is invited to look up the expression "document automation" on the web to delve into greater detail.

obtained from competitive bid submissions, line productivity tracking, project cost accruals and schedule progressing, logistic performance, shipping metrics, material inventory costs, etc., will rapidly amount to valunomic datasets from which real-time comparisons, cost estimate preparations, scope of work estimates, vendor statistics, and cross-pollination effects can be derived. Tracking changes over time becomes inherently possible, without running the risk of losing count of past changes. Conversely, the datum structures (format, presentation, contents) underlying the tabular templates can be utilized within the database to generate datum requests (bid submissions *et al.*), akin to the automatic generation of documents above.

■ From database to browsers. Where a workforce operates from different interface environments (office, service trucks, virtual office, manufacturing floor, on-the-move salespeople, etc.), the issue of connectivity to the database comes to the fore. Even when it is not in play, the reality is that database graphic user interfaces (GUI) are too often anathema to intuitive usage—being *user oppugnant*. The solution is to introduce an additional interface layer between the database and the user, via a web browser acting as portal into the database. Such a portal must be designed SOLELY from the perspective of efficacy and intuitiveness of user interacting with the system. The complexity of the database MUST NOT be visible to the user. Such a design CANNOT be carried out as a programming exercise or an IT project, since these endeavors are historically incapable of proceeding from the user perspective.

■ Digitalized submissions. This solution is enabled by the integration of the previous two solutions. Simply put, all information streams flowing into the organization (endogenously or exogenously) go through the database-specific browser portals. Take the example of an Invitation to Tender (ITT) for the design and construction of an industrial facility. It will comprise several ponderous scope-defining documents, imposed standards, and tabular templates to be used by bidders to provide their answers (and especially their itemized pricing details). Rather than submit their bids by email with document and spreadsheet files, bidders will be uploading their bid contents digitally via the portal. From there, the comparison of the bids will be vastly accelerated by the analytical applications embedded in the dynamic knowledge set database.

■ Digitalized workflows. This solution has been around for years in the corporate world, through modules embedded in ERP systems (others are offered as standalone applications). These applications handle the review and approval of documents electronically, via pre-defined approval hierarchies programmed into the system. These applications offer the seminal advantage of formalizing an organization's approval process and inject blinding visibility into the progress of the process. They are especially potent in networking people sitting disparately within and without the organization. Traceability, audit-proofing, and bottleneck tracking are guaranteed. These applications, however, often suffer from three efficiency-destroying characteristics: 1) they make it too easy to inject dabbling into the review process,* thereby excoriating the principle of *direct accountability* introduced in Chapter 3) they are ill-connected (or not all) to the sources of the supporting documents; and 3) they are frequently *user oppugnant.* Fortunately, redress is readily at hand: dabbling is countered by a pro-ordained approval hierarchy built upon the principle of *direct accountability.* The disconnect is solved by the integration of the plik management system (see fruity solution above), the dynamic knowledge set, and the *virtual kernel* solution (described below). The third drawback is solved either by adding a browser portal in the manner described above or by better customization of the GUI modules included in the workflow software.

■ Virtual kernel. This solution radically changes how an information dataset is associated with a physical asset. This physical asset could be a plant, an installation, or a system; it can be the manufacturing floor and the activities happening on it at any time; it can be a formal project established to pursue the development of the physical asset. The kernel is constructed from a virtual, 3D representation of the physical configuration of the asset.† The model gives the user the ability to "walk" inside the model via a computer display and access whatever piece of information exists for anything visible on the screen (including schematics, dimensions and weights, component datasheets, instruction manuals, operation videos, purchase orders, historical reliability data, past vendors, associated drawings and elemental 3D models, performance simulations, construction specifications, quality control reports, etc.). The virtual physical configuration is the foundation of the model. Every other piece of data, records, knowledge, and intelligence generated over time is tied back to that configuration (through databases integrated

---

* The *dabbler's curse* is explored at length in Chapter 6 of *Investment-Centric Project Management.* It is one of the primary causes of capital projects' consistent failures to come in on time and on budget. The principle of *direct accountability* has its origins in the desire to eliminate dabbling.
† See also the 3D kernel in Chapter 18 of *Investment-Centric Project Management.*

together and operating in the background). The *virtual kernel* comprises the foundational configuration model and all information assets attached to it. Evidently, the kernel will be ever changing in time and evolve dynamically as circumstances change.[*]

■ Geospatial linkages. This solution is a derivative of the *virtual kernel*. To many organizations, the physical location of the data generation sources that are fed back into it is as important as the datum streams themselves. These sources could be static (a plant, a piece of equipment bought by a client and installed somewhere, the location of a weld on a specific joint of a buried pipeline) or dynamic (the whereabouts of a repair crew, the location of a delivery vehicle, the progress of a transmission power line construction crew, the weather conditions). When such data are required, the spacetime details should be compiled automatically (via transponders for example) or by manual uploads via yet another browser portal. Additionally, when pertinent, the spacetime details should be tied back to the *virtual kernel*.

■ Report visualization. Traditional reporting methods have been limited to the graphing capabilities of spreadsheet and database software. Bar charts, trend lines, pretend 3D graphs (adding the illusion of depth), and other flat (i.e., 2D) layouts are the norm. These primitive presentation schemes suffice for the simplest cases but are woefully unsuited to more complex analysis results produced by data mining, neural networks, and AI systems. To put it bluntly, spreadsheets and general-purpose database applications are incapable of conveying these types of results in a clear, concise, and intuitive manner. Visualization also goes beyond result presentations, where dynamic/transient datasets must be viewed as a function of time in conjunction with analytical results (construction progress, traffic patterns, crowd movements, weather patterns).

■ Reality visualization. This solution is the twin of report visualization. It is required when an organization wishes to perform group reviews of virtual models. Consider the example of a facility that is being designed and needing a review of its layout by the design team, the executive team, and the plant operating team. The hick: these three groups are located in three different cities and getting everyone in the same room would be cost prohibitive (and wasteful of precious time due to

---

[*] The concept of the *digital twin* arrived on the scene around 2015 or thereabouts. It shares the notion of the physical configuration and the computer display as an access portal into the asset's *design and construction* information but falls short of the more exhaustive functionalities of the 3D kernel or the virtual kernel considered herein.

flying). The solution is already here, courtesy of virtual reality gaming companies. It is entirely possible to utilize the internet to connect everyone involved with VR displays (supplied locally) and perform "walkthroughs" and "flythroughs" jointly, in real time, from the same 3D model. Other systems going by the moniker "augmented reality" (AR) are also available commercially to increase the depth of possible interactions between the participants and the model.*

# Numer Solutions

## *Could vs Should*

Numer solutions are fundamentally different from their Plik counterparts because of the physical constraints they entail. Plik solutions are rolled out; numer solutions are constructed. This means higher capital and operating costs, and greater systemic complexity. Just because anything and everything in the numer world can be digitalized does not mean that it should. For starters, the sheer volume of numer-generated data will always outweigh their plik counterparts, necessitating an entirely different, and costlier, storage architecture. On the flip side, what matters in numer data is not the preservation of records, as is the case for plik applications; it is the detection of trends and outlier excursions from the norm, which brings up another difference with the plik set: the requirement for continuous datum analysis. The latter in turn implies, *inter alia*, specialized software, which carries additional cost drivers in the form of licensing fees, expert personnel, and dedicated computing hardware. *In fine,* the numer game is played in a radically different realm than the plik's, one that is focused on transforming data into actionable decisions.

---

* VR and AR differ in the way a participant's presence is experienced. VR places the participant at the center of a simulated environment on a computer, without any connection to the room in which the participant stands. The participant is transported in the imagined world. All popular gaming systems are based on this premise (such as Oculus Rift, Samsung Gear VR, Google Cardboard, etc.). AR starts with the participant's surrounding reality and adds details to it through computer imagery. The participant can see the physical surroundings, while virtually imagined details are superimposed on his field of view. Microsoft's Hololens is the best-known commercial system at the moment. VR systems require dedicated visors to be worn; AR systems can utilize a participant's computer screen or even smartphone.

Organizationally, the numer set is operated in an enabling role to business functions, whereas the plik set is itself a standalone business function.

The ramifications of this focus are twofold. First, one must assess the valunomy of digitalizing a given piece of equipment or process. Will the data generated by the digitalization be required for ulterior actions and decisions? If not, digitalization is a waste of time and money. If yes, exactly what data are required to make those decisions and at what frequency must they be produced? What or who must perform what analyses to trigger which decision requirement? Once triggered, what or who makes the decision, and how will it be implemented? These questions hint to the reader at the extent of the planning that must go into the digitalization strategy; we are way beyond the planning involved with the plik set. Fortunately, the answers to these planning questions will be found in the *model calls* formulated in the previous chapter. The matter of continuous datum analysis, for its part, is addressed next.

## Sentience at the Edge

Datum analysis is a function of the position of a numer instance along the *knowledge spectrum*. We ask what actions must be performed on the numer's data? Some will be performed by the numer instance (formatting and transmittal for example). Other actions will be by external agents (averaging, trending, flagging for instance). Next, we ask where each action is best performed, i.e., by numer instance or by one or more external agents? The tool to answer that question is the *nodal structure* (see Figure 3.3). There are no right or wrong ways to decide which level will perform what action. Each choice simply entails specific degrees of costs and complexity. Take for example a sensor meant to monitor the temperature of a heat-generating machine. It could be limited to layer 2 or 3, whereby the temperature is measured, turned into a numeric byte, and transmitted to a remote-control system. Alternatively, it could be equipped with sophisticated software commensurate with a nodal layer 4 or 5, or perform its own assessment of the measured temperature, and relaying the outcome of that assessment back to the machine's controller to alter its running parameters. Either way can be made to work, obviously, but with distinct implementation details and algorithmic complexity. The delegation of the algorithmic functions back to the sensor is called "edge computing", an expression that pervades the IoT world. Absent such delegation, we are left merely with "traditional computing". So, which way to go, edge or traditional?

There are no hard and fast rules. State-of-the-art chipsets have slashed the hardware costs of edge computing to the point where it is economical at all scales. The cost drivers have instead shifted to the lifecycle costs of the software embedded in the chipsets. All things being equal, there is an argument to be made in favor of operating numer equipment at layer 5 or 6, where the equipment achieves a high degree of algorithmic sentience. High sentience yields high degrees of machine autonomy and system reliability. Autonomy in turn reduces the computing and memory demands upon centralized control systems. Reliability, for its part, empowers the organization to implement just-in-time maintenance activities, reduce spare inventory volumes, and monitor equipment performance at the individual level, from which timely TCO analyses can be carried out at will. Another benefit of higher sentience levels is the capability of the organization to operate the sum total of its numer assets as a closed IoT, which could eventually be tied to external IoT networks to yield even more valunomic efficiencies.

The flip side of high levels of numer sentience is higher software maintenance and cooling costs. Sentience being synonymous with advanced algorithms, the issue of firmware becomes preponderant. These kinds of algorithms are anathema to tinkering by lay technicians. The operator becomes effectively beholden to the *creators* of these algorithms. Firmware comes shackled with perennial service contracts. Firmware updates can become a nightmare when required to be deployed simultaneously across hundreds or thousands or more numer devices—with the dreaded potential for system failures, hiccups, or data losses.

Intra-device compatibility is another crack lying hidden in the digital construct, especially when the organization operates multiple brands of numer devices from myriad vendors. It is not enough for devices to be individually complete; they must also be able to play and communicate with each other. Who will make sure that they can? If they cannot, which ones are to blame? This question remains unsolved by the industry at this time.

Along with compatibility comes the issue of standard proliferation, which is embedded into the very fabric of algorithms. Unless these standards are settled (and frankly, today, they are not, nor will they be in the foreseeable future), the organization runs the risk of adopting orphaned standards. There are more potential dangers at higher network levels, where numer devices are operated as symbiotic/holistic systems. Strange coupled behaviors could arise unbeknownst to the operators, nor expected by designers—very much akin to the mysterious ways by which artificial intelligence algorithms arrive at results without anybody understanding how they got there. This issue of dynamic coupling (in a mathematical sense) is only now beginning to appear on the radar of algorithm designers, mainly because the technology is still nascent and has yet to reach the market penetration necessary to give form to those behaviors. But it is bound to dominate the discussion once IoT has gone mainstream.

## Internet of Things

Despite its ubiquity in the public sphere, the concept of an internet of things is mistakenly understood to mean machines of the world, unite! In fact, any organization can operate its own, proprietary internet of things. All that is needed is a series of numer instances operating at nodal layer 3 or higher, connected in such a way that they work together without human interactions. Therefore, from a design and engineering standpoint, an IoT is no more mysterious than any plantwide control system. Differences (and benefits) begin to appear when the networked numer instances are endowed with higher layers (4, 5, or 6). Then, the network is conceptually able to monitor its own performance and alert its operators of potential sources of disruption to its normal operating envelope. The network can also take over commercial transactions on its own, based on predefined *go-no-go* criteria, without the need for a human to plod through friction-inducing procurement procedures.

The principal areas of concern, for the organization, do not lie in the technical realm *per se*, but at the governance level, with the keyword being "integrity". The IoT must possess five types of integrity: technical, operational, security, compliance, and transparency.

- *Technical integrity* pertains to the physical and interfacial harmony between interconnected numer instances. Physically, the instances must "bolt-up" seamlessly, without interference or incongruity (something that should be self-evident). Interfacial harmony boils down to communication: interconnected numer instances must speak the same language using the same syntax, structure, formats, and definitions (another self-evident truth). Algorithmic integrity alludes to the backward and forward compatibility that is required by the software embedded into a numer instance. Software updates must be consistent with past versions of the same instance AND with the input/output characteristics of competing software, loaded into commercially distinct instances that are nevertheless connected to this instance on the IoT network. Bugs and incompatibilities cannot be permitted to disrupt the seamless operation of the IoT.

■ *Operational integrity* pertains to the ability of the network to operate without unplanned interruptions when updates, upgrades, and other like changes are made to it. Fundamental to it is requirement for *algorithmic integrity* among the interconnected numer instances. The expression alludes to the backward and forward compatibility that is required by the software embedded into a numer instance. Software updates must be consistent with past versions of the same instance AND with the input/output characteristics of competing software loaded into commercially distinct instances that are nevertheless connected to this instance on the IoT network. Bugs and incompatibilities cannot be permitted to disrupt the seamless operation of the IoT.

■ *Security integrity* alludes to the imperative for the IoT network to be impervious to external threats from hackers, disruptors, spies, and uninvited "visitors". Security integrity is by far the weakest link in any IoT network, owing mainly to the near absence of encoded security protocols at the level of the numer instance. Billions of sensors and devices the world over are deployed across myriad networks, IoT and other wise, without a shred of protection against nefarious incursions.

■ *Compliance integrity* refers to the legalistic frameworks being deployed by nations and institutions worldwide to protect user data, enforce data privacy, and govern data ownership. Compliance integrity goes hand in hand with security integrity; the former takes center stage once the latter is found breached. Compliance integrity must be designed into the digital construct as the cornerstone of its functional requirements and design specifications. One does not build compliance integrity into a network; rather, one builds a network upon a compliance integrity foundation.

■ *Transparency integrity* refers to the importance of an IoT's operator to be able to have complete visibility into the inner workings of the network, especially when things go awry or produce an unexpected response. Without transparency, operators will not be able to troubleshoot issues, preempt security threats, or understand how a particular decision or outcome of the network was modulated. Transparency becomes much more important when layers 6 and 7 are present in some of the numer instances: artificial intelligence is notoriously opaque on its neural meanderings behind a given inference.

## *Data Lakes*

Numer devices will produce a huge amount of data, possibly transacted and stored across a multitude of platforms. The binary firm will end up with a mixed bag datum sources from traditional enterprise activities, machines and sensors, IoT, and social media. These sources yield myriad datum types (unstructured and multi-structured), across several interface types and latencies from batch to real time. We will speak of *data warehousing* when data are stored and archived in a structured, ordered fashion. Such a hierarchical approach benefits the visibility of the digital assets, but at the price of slow analytical speeds. With *data lakes*, the structure is jettisoned; all data coming into a storage location unstructured, without any processing. Such raw data are far more agile than their warehoused counterparts, because they can be transformed at will to fit the particular needs of an analysis. This big free-for-all character benefits the speed of analysis but comes at the price of security vulnerabilities. Managing these sources and extracting valunomic intelligence from them requires a unifying architecture paradigm that recognizes the impossibility of doing all this from a single data storage platform. The paradigm calls for the employment of a portfolio of platforms that can be managed at the interfaces. According to Philip Russom,* from the consultancy TWDI, the platform portfolio becomes a corporate data management strategy calling forth several data platforms which hold data distributed physically across numerous databases and storage sites. Russom defines multiplatform data architecture as a variegated mixture of old and new data, managed centrally or from the Cloud, with an array of software solutions woven together upon a distributed data architecture. This is no longer the quaint land of simple client/server architecture that prevailed in years past. They are aggregate data management super systems characterized by unbridled diversity, extreme sophistication, and uber complexity.

An alternative to handling these complex issues internally is to outsource it. This is what Bernard Marr, in *Data Strategy,*† calls BDaaS (big data as a service). BDaaS levels the playing field for the small organization, who suddenly gains access to a world of high potency analytics costing a fortune. The organization no longer owns the digital construct: it simply hires it from a third party. The field is already crowded by the behemoths of this world (Amazon, Microsoft, IBM, HP), thereby ensuring stability and longevity to

---

* Philip Russom: *Multiplatform Data Architectures: Addressing Complex Modern Requirements for Hybrid Data and Its Business Use.* Best Practices Report Q3-2018, by TDWI.
† Bernard Marr. 2017. *Data Strategy.* Kogan Page.

the relationship. BDaaS, however, comes with a caveat: it is phenomenal in uncovering trends in sales, markets, and customers, but nearly impotent to the pursuit of organizational efficiencies. For the latter, a dedicated in-house digital construct is necessary.

## *Business Intelligence*

The majority of the *model calls* assigned to the commercial performance of the organization can be addressed by commercial software solutions that are already established in the marketplace. The buyer is faced with a surfeit of viable options tailorable to the organization's unique requirements. A discussion of the pros and cons of the marketplace's wide array of choices is beyond the remit of this book. The reader can get started down the evaluation path with the following suggestions: Birst, Board International, Domo, IBM Cognos, Information Builders, Izenda, Logi Analytics, Looker, Microsoft, MicroStrategy, Oracle, Panorama, Pyramids Analytics, Qlik, SalesForce, SAP Business Objects, SAS, Sisense, Statsbot, Tableau, Targit, ThougthSpot, TIBCO Spotfire, YellowFin, WebFOCUS, White Whale Analytics.

The list (presented alphabetically) is neither comprehensive nor suggestive of endorsement by this text; it is merely presented as a snapshot of what is available to the reader *today*.

# Distributed Ledger Technology

## *The Big Deal with DLT*

Let us begin with a caveat: cryptocurrencies (bitcoin, Ethereum, the rest) do not belong in a digital management transformation. They are not the killer apps of the DLT world and remain mired in a land of confusion in regulatory and legal terms. The hype is simply too big for the stage. DLT, on the other hand, is advancing incrementally worldwide as a feasible technology (albeit not quite in the disruptive manner that the hype also purported). It is a technology that has a place in a digital transformation, provided that it is deployed intelligently. For all DLT applications are variations on the database theme. The traditional database framework is still the dominant paradigm to most organizations. Generally speaking, it is the superior solution over a DLT implementation.

When, then, is DLT the better solution? When the following characteristics of the organization's holobiome are in play:

■ *Mistrust of holobiome players/partners/vendors.* This is a harsh word from which the reader may recoil at first sight. Nevertheless, mistrust is at the heart of all interactions between organizations. That is why written contracts were invented, and why the choice of a legal jurisdiction to arbitrate disputes is so important.* Yet, even within a stalwart legal environment, mistrust may remain in play—having a contract is no guarantee that all parties will willingly live up to all its terms. DLT becomes a potent tool to forcefully apply the integrity criterion to the transactions in play. There is no need of a contract when trust reigns.

■ *Opacity of transactions.* If mistrust taints the integrity of the information transacted, or the processes applied by the parties, the contract becomes nearly futile as a guarantor of the mutually satisfaction of the parties involved. For instance, anybody can draft up a quality assurance that looks good and credible. But its contents may be specious, even massaged to hide troubles. Proof of origin is another point of contention when dealing with a global supply chain. Drug manufacturing, steel making, casting porosity, and active ingredients in a chemical are all examples of potentially flawed products derived from dubious supply sources and/or manufacturing processes. Their true nature is opaque to the recipient. DLT can inject the needed transparency throughout the entire supply chain.

■ *Contract monopsony.* Although the parties to a contract are deemed equal in a court of law, the truth is that the one holding the money tends to have the upper hand when contractual tergiversations appear. A vendor that delivers a shipment on time and in compliance with all quality assurance demands could still end up getting paid 90 or even 120 days later than the payment clause spelled out in the contract. DLT are uniquely capable of obliterating this monopsony proclivity through the use of "smart contracts" equipped with actionable triggers when preapproved criteria are met by both parties to govern the reaction of one in response to the action of the other.

---

* If given the choice between London, Paris, New York, or Tokyo as the jurisdiction of choice, logistics, convenience, and costs will influence the decision. Now, throw in Beijing; unless one of the parties signatory to the contract is a dominant Chinese company, nobody will choose Beijing, even if they happen to be closer to it than the other locations. There lays the importance of the jurisdiction.

■ *Dabblers, meddlers, and middlemen.* Global commerce is riddled with friction-inducing intermediaries who add nothing but costs to a transaction between a buyer and a seller. Some of this dabbling is legal, some of it is pure graft, blatant or disguised. A DLT framework is supremely potent in eviscerating costly dabblers and eliminating transactional friction between buyers and sellers.

Ultimately, the distributed ledger technology enables two or more distrustful parties to transact among themselves through the creation of permanent, unalterable records, visible unconditionally to all parties, and without having recourse to a central/meddling authority.

## Levels of Decentralization

DLTs come in a one of four forms, ranging from entirely private to entirely public. The DLT first form is the traditional database, controlled centrally and uniquely by the organization. *It is not economically feasible to transform a centralized database into a DLT.* The foundation, the plumbing, the DLT transaction rules, and the types of records to be maintained must be designed from the ground up and rolled out uniformly among the parties privy to the DLT framework.

The DLT second form is *permissioned, private, and system shared.* This will be the default setup for most organizations. One classic example is an oil and gas pipeline operator with hundreds or thousands of kilometers of buried lines. These pipelines are welded together from standard length pipe joints. Each joint is manufactured in a batch, against exacting chemical specifications for the steel alloy. Every joint is uniquely identified, uniquely documented, uniquely welded and tested, and uniquely monitored when in service. An enormous amount of data is therefore generated by an army of suppliers, shippers, inspectors, constructors, and testers, between the time the pipe is ordered until it is buried and flowing with the revenue-generating fluid. A DLT second form is ideally suited to bring order to the chaos stemming from the mounds of information accumulated. It is private, to the extent that only those parties supplying information to the pipeline operator can participate in the DLT. It is permissioned to control who is on the DLT invitation list. And it is system shared because all invited participants are required to utilize the same software and protocols to interact with the DLT framework.

The DLT third form expands the franchise to the public (usually in limited fashion). Our pipeline operator will be governed by integrity rules imposed by the pertinent regulator. Some of these rules may require the operator to publish in the public domain the results of the recurring pipeline inspections. The DLT second form can be modified into the third form by allowing the public to view the inspection results in real time (but limit the access of the public to recording data to the ledger).

The DLT fourth form is best illustrated by bitcoin. It is everywhere all over the world. It is also the least desirable from an organizational perspective. Only the very largest corporations operating globally will have a need to get into this fourth form. For everyone else, it is best to be no higher than the third form.

## Dangers and Pitfalls

DLT are not the panacea that some paint them to be. First, a DLT is effectively an operational platform (akin to an operating system for a laptop). At the moment, the world is dominated by three major platforms: bitcoin, Ethereum, and Hyperledger, followed by a plethora of smaller offerings geared toward niche applications. Each platform has its strengths and its weaknesses, and none is able to operate as an all-encompassing generic platform suitable for all purposes.

Second, none of these platforms play well with others. Hence, to adopt one platform is to isolate oneself from all others. Each platform is an application orphan plagued with greater odds of not surviving the passage of time (imagine the tribulations of Microsoft vs Apple's operating systems, multiplied by a thousand). Unless your organization is actively working on its own proprietary platform development, it is best to stick with the biggest names at the moment and let others sail these turbulent waters

Third, the DLT industry remains scattered and unseasoned overall. Few cities in the world can boast to possess an established base of proven experts. For most organizations, this means a dearth of competent, available consultants to draw upon to embark on the DLT bandwagon. Furthermore, the incessant jostling of aspiring platforms, rolled out sooner than their underpinning technology maturity, brings about failures that destroy the public's frail trust in the technology. Outside of the finance sector, the realm is a jungle.

Finally, the very nature of DLT flies in the face of old, staid regulatory and legal frameworks that will not abide by the decentralizing features of this technology. Remember that DLT are, at heart, nothing but sophisticated algorithms, capable of illustrious programming and daft coding blunders all at the same time. Yes, smart contracts can be hard coded into a DLT, but nothing stops these contracts from being erroneously coded with ersatz "smart" conditions that could blow up in everyone's faces.

## *This One Can Wait*

DLT is the only binary opportunity in this book that is NOT endorsed for adoption. The globally disruptive, world-changing revolution promised hitherto by DLT has not come to pass. The cryptocurrency mania ushered in by bitcoin, hailed as harbinger of a brave new banker-less world came crashing down in 2018, along with the collapse of that particular DLT segment. The hype blew up then blew out. According to McKinsey, a consultancy, the technology has utterly failed to deliver on the expectations. Nothing substantial percolates to the surface, beyond the ceaseless promises of the technology's proponents. McKinsey's report bluntly dismissed DLT as a viable technology. Whatever problem DLT systems claim to solve, there always seems to be a simpler, cheaper commercial solution available now.*

HOWEVER, if you still wish to get on the DLT bandwagon, the best way to go about it is from the perspective of risk mitigation.

The safest and easiest way to dip one's toe into the water is as a supplier/vendor to a client with a presence in a working DLT environment. Volunteer your organization to partake into the DLT, within the scope of your information demands. Assemble a small DLT team whose mandate will be to acquire, deploy, and implement the interaction protocols, which will, perforce, be limited in the initial stages (you want limited, by the way). Your team learns the ropes on the fly, through its exchanges with other DLT teams from without. Your DLT team will be focused on the mechanics and mechanisms of datum transactions on the network. A second team will be necessary to bridge the needs of the DLT team and the datum production of the organization. Fortuitously, this second team is already in play: it is none other than the digital management transformation team! Most likely, the information needs will entail changes to the way business is conducted from within, along the kinds of transformations discussed hitherto for the *plik* and *numer* sets.

---

* See McKinsey & Company's January 2019 report *Blockchain's Occam Problem*. Available at: https://www.mckinsey.com/industries/financial-services/our-insights/blockchains-occam-problem.

In this first incarnation, your organization's role will be limited to information sourcing only. The next step up from there is to become an active DLT participant, by taking on "miner" duties. A miner gets involved with the mechanics of proofing a new transaction occurring on the DLT, with majority validation of the transaction, and with the creation of actual transaction records on the ledger itself (on which new "blocks" are irreversibly added into the chain). The second level is also where the organization can begin to exploit its own IoT network.

For most organizations, and indeed the vast majority of them, the second level of DLT incarnation will be their last. The next level up takes the organization into the design and development weeds of DLT platforms. This level entails an entirely different skillset by the DLT team, one that is not required at lower levels. We are now speaking of DLT designers and engineers, a domain that exceeds the remit of this book.

# Intelligence Solutions

## *The Big Deal about Big Data*

The plik and numer solutions cater to datum transactions in real time. The visibility is experienced at the level of direct usage but lacks the ability to elevate the perspective into the ecosystem in which those transactions take place. For some organizations, these gains suffice to justify their binary transformation; but not for others keen on understanding how everything really works organically. This is where intelligence solutions come in, and the point of entry into the realm of business analytics, where big data, machine learning, neural networks, and artificial intelligence roam. These so-called *intelligence solutions* move the organization toward the extreme right of the knowledge spectrum. While the gains should be expected to be huge, the TCO of those solutions can become giant hurdles to overcome and subsidize. Which begs the question: When is it worth it to go there? A question whose answer is formulated into a W5H argument:

- WHY: because the organization lacks the knowledge and understanding of its information assets to manage them effectively.
- WHAT: to enable data discovery and cataloging, information and knowledge profiling, datum lifecycle, trends and insights, inferences and connections.

- WHO: chief data officer, chief information officer, IT director, data security managers, consultants.
- WHERE: outsourced (initially) or in-house (eventually).
- HOW: data mining, machine learning, neural networks, artificial intelligence.

Pragmatically, getting into these advanced solutions is justified on the basis of improving the organization's visibility into its information assets; enhancing datum transaction fluidity (reduction or elimination of friction); and producing actionable insights that can drive business performance. The visibility vector warrants emphasis over the sexier analytical promise. According to IDC, a consultancy, organizations will spend 81% of their datum transactions on custodial tasks (search, prepare, protect), and only 19% on genuine analytical tasks. This 81% amounts to profit-killing operational expenses. *Intelligence solutions* can flip this ratio, thereby eliminating 60% of the costs of the custodial tasks.* In a sense, the unheralded valunomy of *intelligent solutions* lies not so much in their vaunted analytical prowess, but in their innate ability to wrestle custodial tasks into the bottom-line ground.

## *A Fickle Proposition*

It may come as a surprise to some readers that AI is already a mature technology, albeit not so ubiquitous as it could be. The worldwide brouhaha about data privacy and ownership is evidence of it; they are a thing because they feed AI applications. The lack of ubiquity can be blamed on the complexity and costs of AI technologies. Hitherto, they are to businesses what fighter jets are to private plane owners. AI is on the rise, indubitably, but lingers at the horizon. Getting into the game is worth the effort, regardless, but entails three immediate challenges, according to Bernard Marr:

- Challenge 1. It's not about large datasets but about the right set of data and about having the right to hold and manipulate the data. AI applications are arrantly specialized to work on highly specific datasets. There is no such thing as a universal AI.
- Challenge 2. One platform to rule them all. AI cannot work effectively when datasets are spread wide across disconnected data islands and

---

* IDC PlanScape: *Data Intelligence Software for Data Governance.* International Data Corporation, Report WP248-En, 2018.

entry portals running their own agendas. Organizations must corral their information assets and herd them toward a central point from which AI functions can be performed.

■ Challenge 3. AI is a compound specialization. You cannot improvise your way into AI. There are no "AI for Dummies" books to make you into a coherent, competent AI operator. There are no generic recipes and lists of nine, ten, or eleven secrets to a successful AI capability. You will need data scientists who understand the physics of datum analysis, visualization experts who can engineer intelligent presentation schemas, and system integrators who can design the plumbing to network all datum sources into the single platform called by challenge 2. Finally, organizations must educate their workforce on the ramifications of the AI paradigm.

To these three challenges, we can add four more:

■ Challenge 4. AI isn't IT. The gravest organizational mistake is to think of AI as something that belongs to the IT department. AI will, obviously, rely on the information infrastructure managed by the IT department, but it will operate entirely separately as its own department or division. AI is not an IT support function, it is a business function.

■ Challenge 5. AI is about people. In the context of a corporate strategy, intelligence solutions are ultimately about people and their interactions with data, rather than being about high-tech software, powerful machines, and immense data servers. While it is true that AI will inevitably displace people in many traditional roles, there will never be an organization that runs purely on AI. People are still running the show and ruling the AI infrastructure. What AI will do is enhance people's decision-making capabilities.

■ Challenge 6. AI ain't cheap. That much should be abundantly clear to the reader by now. At this point in time, everything associated with AI is costly: the people, the applications, the hardware, and the data governance enabling it. Nor is it possible to save money by going cheap wherever possible: going cheap means an *incapable* AI framework. The easiest way into AI is to outsource it, lock, stock, and barrel. The next step is to build the single platform infrastructure needed to create, capture, store, and manage the datasets, but outsource the AI performance. The third level is to bring this performance in-house.

■ Challenge 7. Bias vigilance. Systems that require training to recognize patterns are notoriously sensitive to unintended biases, right down to the raw data that power them. Fantastic prediction blunders have made the news over the years and will continue to plague the technology's reputation. Datasets that are homogenous will be useless to make predictions beyond that homogeneity. Narrow sets of data produce results deemed representative of larger system behavior, while being inherently biased (something called *availability bias*). Another one, *confirmation bias*, results from filtering out data that did fit the expectations. Biased results may not appear biased and so create the illusion of result correctness, from which decision blunders could follow.

Never ever entertain the notion of developing your own customized AI applications, unless that is the core business you wish to pursue. AI is like desktop apps times a million in complexity.

## FERAL MACHINES

For all their wondrous abilities, AI machines cannot be trusted with wanton abandon, quite the contrary. There is a malicious irony with the fact that humans have mastered the art of designing neural algorithms but have absolutely no clue as to the actual deep logic by which these algorithms arrive at their results. Indeed, at least for the time being, there exists no method to test and validate the *behavior* of neural networks, either in real time or *post mortem*. It would seem that AI works in mysterious ways to the human observer.

Compounding the impenetrable opacity of the inner workings of AI are two other character traits of AI algorithms that erode one's trust toward them. The first is the—surprising some may say—severe lack of adaptability of AI systems. They work wonders when they are subject to highly calibrated, laboratory-level conditions. Change a few pixels, literally, and the inference can yield laughably wrong results. The second trait is vulnerability. Because AI systems are so arrantly beholden to their training sets, they become by the same token exceptionally susceptible to hacking and pirating by external actors. A competent nefarious hacker has no need to infiltrate the neural algorithm of a given AI. His task is rendered embarrassingly easy by meddling instead with the input data stream that is being fed to the AI. In a real-time application (say, crowd traffic in an airport), the hacker needs only pester the data stream, which is sadly too easy in this day and age (see *security integrity* above), to rattle the AI's algorithmic protocols to produce wrong inferences.

## The Opportunity Domains

The decision to adopt an *intelligence solution* must be driven by the *model calls.* By opposition, one does not start with AI then figure out what to do with it. The nature of the model calls pretty well dictates what will be required in terms of datasets, analytical algorithms, and result presentation. The types of algorithms can be divided into three domains:

■ Specialized AI, focused on narrowly defined tasks, like machine vision.
■ Machine learning, a subset of AI focused on training decision algorithms (neural networks) through the use of large datasets.
■ Deep learning, a subset of machine learning relying on layered networks to achieve exceptional AI performance from even larger datasets.

A fourth domain exists beneath them as well: data science. Data science is an emerging field in theoretical and applied informatics, in which scientists, mathematicians, data and hardware engineering, programmers, and visualization experts work to design the building blocks of AI.

As we will see below, implementing a successful *intelligence solution* requires a narrow focus on the types of questions to be asked. The idea of a general-purpose artificial intelligence system that can address all kinds of *model calls* is a chimera. One must therefore choose judiciously where *intelligent solutions* will be deployed. Looking at the issue from the perspective of analytical demands leads to the partition of the *model calls* into three types of intelligence analyses:

Type A: Exogenous analytics

- Delivering personalized offers.
- Creating personalized campaigns (marketing, ads, etc.).
- Assessing the efficacy of those campaigns.
- Engage/re-engage customers.
- Understand customers and discover customer insights/trends.

Type B: Revenue analytics

- Identify new potential service and product offerings.
- Identify new potential markets.
- Identify cross-pollination and cross-sale opportunities.
- Identify new supply chain partners associated with new potentials.
- Identify supply chain trends.

Type C: Profitability analytics

- Assess the fluidity of data transactions across business processes and functions.
- Assess the bottlenecks, seen and hidden, across business operations.
- Detect trends, threats, and redundancies carried covertly by operational data.
- Measure the aggregate productivity of interdependent business functions.
- Perform real-time analysis of logistics activities and supply chain performance.
- Capture the holistic picture of the organization's human skillsets and shortcomings.
- Improve manufacturing processes and procedures.
- Predict equipment failures, system disruptions, operational overloads, and maintenance shortcomings.
- Infer failure-inducing patterns of hitherto disparate systems and processes.
- Estimate cost escalations in capex and opex budgets, their causes, and their mitigations.
- Quantify future efficiency gains from organization restructuring or forced technology obsolescence.

■ This raises three challenges for heritage players, who must find a way to innovate more quickly while also ensuring their approach is sustainable.

The list is not exhaustive but suffices to give the reader an idea of where intelligent solutions can be brought to bear to deliver immediate benefits to the bottom line. Some of these applications can, in practice, be folded under a single analytical focus, which reduces the number (and cost) of *intelligence solutions* to be deployed. Others will have to be handled individually, on account of the unique features of the datasets underlying them.

## *Preparing the Grounds*

The implementation of an intelligence solution requires extensive planning. The newcomer to the field will encounter a cornucopia of novel expressions that are not at all evident (deep learning, data science, data lakes, backpropagation, classification models, convolutional neural networks, edge computing, reinforcement learning, antagonist networks, humans-in-the-loop systems, dirty data, to name but a few). Programming languages are also plentiful in the marketplace, each one with its niche applications and ardent disciples, with names like Tableau, Unix shell/awk, Amazon AWS, TensorFlow, Jupyter, R, and Python. Each one comes with its own pointed expertise to understand. Clearly, from the perspective of digital management transformations, becoming an expert is not in the cards. The expertise can be hired, at any rate. The transformational concerns lie elsewhere, which is where decision-makers must gain at least a working knowledge. The concerns include:

■ Dirty data (contents or formats prevent them from being fed as inputs).
■ Lack of data science talent: obvious, since the field is commercially still in its infancy.
■ Lack of management support: also obvious, since not knowing about the valunomy of the solutions makes it hard to convince executives to spend the necessary money.
■ Lack of clear questions: the correctness of an AI analysis is intimately conditioned by the precision of the question. Fortunately, we have the *model calls* to get started on the right foot. One still needs the data scientists, however, to refine the questions.
■ Accessibility: clearly, if the datasets are not available, or difficult to access, the potency of the *intelligence solution* will be emaciated.

- Data literacy: it behooves the organization to teach its workforce the AI birds-and-the bees, especially the decision-makers who will be called to rely upon the AI results to stear the ship.
- Diffidence of the decision-makers: AI will rapidly loose its appeal, and its funding, if decision-makers are unwilling to follow through on the knowledge and insights inferred by AI. The solution to this one is data literacy.

Proceeding in incremental steps is crucial to the success of the implementation. The golden rule: start small, with a very simple *model call*, a single enquiry question, and a smaller dataset. The second step is to set up a small team. *Do not form this team as a stand-alone project team left to its own devices to work things out.* The team must be embedded into the functional group within the organization that owns the *model call*. This way, the team will be forced to engage with the many stakeholders (managers, workers, IT support, customers, etc.) who partake in the generation of the dataset and become crystal clear about what questions they are intent on answering. The third step is to begin compiling and massaging the dataset, with constant vigilance for creeping biases. Next comes small scale test runs conducted from the simplest of questions. With each test run, feed the results back into the system and increase the complexity of the question. Check for hidden biases in the results and tweak the system to eliminate them. Step 5: rerun the successful tests to check if they arrive at the same answers. If not, the solution needs more learning sessions. Repeat step 5 until the system achieves a high repeatability of correct answers (at least 90%), at which point the solution is ready to be tested on an actual *model call*. At this time, run a parallel analysis based on whatever traditional method was used by the organization before, and compare the answers from both parties. If they are similar, the *intelligent solution* is ready for a formal presentation to management. Otherwise, the two parties (the AI team and the traditional analysis team) need to discuss the results and arrive at a consensus on which answer is best. Egos and job preservation proclivities will figure prominently in the background while the debate rages on. Hence, the discussion must be chaired by a senior manager who would have been the logical recipient of the recommended answer (from which a decision would be made).

When the discussion ends up favoring the *intelligence solution*, the system is ready to be formally presented to management. The presentation will focus on the nature of the questions asked, the dataset assembled for the test, the answer obtained, and the reasons for believing the answer. The presentation must include a valunomic assessment of the system, the expected ROI from rolling out the system throughout the organization, and the plan for moving forward with its implementation.

## MASTERS OF CHAOS

Lest the reader be left with a negative bias against artificial intelligence, here is an example from the world of physics, courtesy of senior chaos theorist Edward Ott from the University of Maryland. Ott's team succeeded in simulating the first 12 seconds of the evolution of a flame with high fidelity against the reality of it. Twelve seconds may not sound like much, except that, prior to these results, the most powerful computer systems in the world could only manage up to two seconds, beyond which dynamic chaos wrecked any resemblance of the results with the real thing. Ott and his four collaborators managed this feat by utilizing a niche machine-learning algorithm called *reservoir computing* to replicate the underlying mechanics of the chaotic nature of the flame,* without solving explicitly the complex set of partial differential equations that govern the physics in play. The algorithm proved immensely efficient in predicting the onset of chaos and its temporal evolution by exploiting the very chaotic nature of the algorithm itself. Although the reasons why reservoir computing is so good at handling chaos is not yet fully understood, the fact remains that this niche application of AI is now seen as a powerful and reliable tool for modeling the hitherto inscrutable inner workings of a chaotic system. A simply stunning discovery.

---

* The flame model, known in academic circles as the K-S model, was developed in the 1960s by Yoshiki Kuramoto and Gergory Sivashinsky. Even though the mathematical description is perfectly deterministic, the behavior of the solutions is innately chaotic. Chaos carries its own mathematical definition, through an extreme sensitivity of a mathematically coherent dynamic system to initial conditions. A chaotic system will experience wide, wild variations in outputs when tiny variations to its inputs appear. A stable system does not. That is why weather predictions lose all credibility after just a few days: the underlying mathematics are prone to dynamic chaos. At least until someone puts it under the thumb of reservoir computing! See: Wolchover, Natalie. Machine learning's 'amazing' ability to predict chaos. *Quanta Magazine*. 2018. Available at: https://www.quantamagazine.org/machine-learnings-amazing-ability-to-predict-chaos-20180418.

# Bibliography

Cann, Geoffrey; Goydan, Rachael. *Bits, Bytes, and Barrels: The Digital Transformation of Oil and Gas.* MadCann Press, 2019, 209 pages.

Finlay, Steven. *Artificial Intelligence and Machine Learning for Business: A No-Nonsense Guide to Data Driven Technologies.* Relativistic Books, UK, 2017, 150 pages.

Higginson, Matt; Nadeau, Marie-Claude; Rajgopal, Kausik. *Blockchain's Occam Problem.* McKinsey & Company, January 2019. https://www.mckinsey.com/industries/financial-services/our-insights/blockchains-occam-problem.

Keays, Steven J. *Investment-Centric Project Management: Advanced Strategies for Developing and Executing Successful Capital Projects.* J. Ross Publishing, FL, 2017, 419 pages.

Keays, Steven J. *Investment-Centric Innovation Project Management: Winning the New Product Development Game.* J. Ross Publishing, FL, 2018, 309 pages.

Kelsey, Todd. *Surfing the Tsunami – An Introduction to Artificial Intelligence and Options for Responding.* Todd Kelsey Publisher, 2018, 188 pages.

Kranz, Maciej. *Building the Internet of Things – Implement New Business Models, Disrupt Competitors, Transform Your Industry.* John Wiley & Sons, Hoboken, NJ, 2016, 272 pages.

Mar, Bernard. *Data Strategy: How to Profit from a World of Big Data, Analytics and the Internet of Things.* Kogan Page Limited, London, 2017, 200 pages.

McLuhan, Marshall. *Understanding Media: The Extensions of Man.* MIT Press, 1994.

Rifkin, Jeremy. *The Third Industrial Revolution: How Lateral Power is Transforming Energy, the Economy, and the World.* St-Palgrave MacMillan Publishers, NY, 2011, 304 pages.

Rogers, David L. *The Digital Transformation Playbook.* Columbia Business School Publishing, NY, 2016, 278 pages.

Rossman, John. *The Amazon Way on IoT: 10 Principles for Every Leader from the World's Leading Internet of Things Strategies.* Clyde Hill Publishing, Washington, DC, 2016, 168 pages.

Russell, Stuart. *Artificial Intelligence – A Modern Approach.* Pearson Education Limited, London, 2015, 1164 pages.

Sacolick, Isaac. *Driving Digital – The Leader's Guide to Business Transformation Through Technology.* American Management Association, 2017, 283 pages.

Schwab, Klaus. *The Fourth Industrial Revolution.* Crown Business, NY, 2017, 192 pages.

Tapscott, Don; Tapscott, Alex. *Blockchain Revolution: How the Technology Behind Bitcoin Is Changing Money, Business, and the World.* Portfolio/Penguin, 2016, 432 pages.

Vigna, Paul; Casey, Michael J. *The Age of Cryptocurrency: How Bitcoin and the Blockchain Are Challenging the Global Economic Order.* Picador, NY, 384 pages.

Westerman, George; Bonnet, Didier; McAfee, Andrew. *Leading Digital – Turning Technology into Business Transformation.* Harvard Business School Publishing, 2014, 256 pages.

Windpassinger, Nicholas. *Internet of Things: Digitize or Die: Transform Your Organization: Embrace the Digital Evolution: Rise above the Competition.* IoT Hub Publishers, NY, 2017, 284 pages.

# Chapter 7

# Engagement

*The journey of a thousand steps... has begun.*

Part IV of the digital management transformation puts the rubber on the road. We move from planning and decisions to plans and implementation. By the end of this chapter, the organization will be operating digitally.

## Implementation Strategy

The previous three chapters helped define, quantify, and select the many components of the organization's bespoke digital transformation. In this chapter, the elements are integrated into an operational design primed for roll-out across the organization. The basic approach to implementing this design is the same regardless of the extent of the organization. The first step is the actual integration of the elements into the *digital construct* (hardware, software, tools, techniques, processes, and procedures), as well as the plan to purchase, assemble, install, test, and start it all up. The second step addresses the question of how the organization will manage its newly minted digital operations, what we will call corporate digital governance. As the name implies, governance erects a formal framework populated by policies, procedures, and reporting lines to manage the functions of the digital construct. It is common to carry out steps 1 and 2 in parallel, to save time and money, and insert a feedback loop between the two for the purpose of fine tuning the governance details.

The third step rolls out the digital construct and the governance schema operationally, in a limited scale. One functional area of the organization is chosen for this pilot implementation. Note that the digital construct will only be constructed in part, to the extent required by the *model calls* assigned to the pilot group (for instance, the chosen group may not be tied into an internal IoT schema). During the pilot implementation, things get powered up in an electrical sense—hardware, software, and devices get energized and begin producing the outputs for which they were bought. Obviously, the pilot implementation is likely to be plagued by bugs, inconsistencies, disconnects, gadget failures, application crashes, and datum losses. All these setbacks are normal for this kind of complicated system integration. Even elements of governance may need revising, once reality obliterates the hidden deficiencies of idealistic prescriptions. The expectation of multitudinous problems bears heavily on the choice of the pilot group (not unlike a guinea pig). There may be more than one iteration of the pilot implementation as well, according what has been missed in the planning of the implementation. It is for that reason that the reader is urged repetitively to proceed incrementally during the pilot phase. It is easier (AND faster AND cheaper) to fix bugs and problems a few at a time. Going incremental is the safest way to manage and corral the implementation risks. Or, to echo a piece of wisdom by Margaret Keays: *Make small happen.*

Note that the human element of the operational digital construct will reveal, rather quickly, what training is required to get people up to speed in working in this digital paradigm. This will be especially true of the training that was initially dispensed, whose contents may not have anticipated all user needs during their development.

Finally, we come to step 4, which should begin—very—soon after the pilot implementation has been completed, and its learnings thoroughly assimilated by the transformation team. Let us emphasize the benefits of proceeding incrementally. The functional areas of the organization should be transformed one at a time by the roll-out crew *augmented with the pilot crew and the input of an area's management team.* The order of things unfolds sequentially in four stages:

■ Stage 1: Consider each functional area as a network node characterized with its own set of *model calls* and *endogenous* datum streams (the information requirements). Each node operates as a *system*, in keeping with the concept of functional levels in Chapter 5 (under the heading "endogenous model calls" and note 6). Make sure that all inner workings are indeed working well.

■ Stage 2: Integrate the areas that are directly connected by their exogenous information requirements. Areas so connected form a primary *nodal installation* (again pursuant to Chapter 5). Once again, make sure that the nodal connections are validated, verified, and proven to work seamlessly together.

■ Stage 3: Integrate the primary *nodal installations* into a wider, secondary *nodal installation*, through their exogenous information requirements. At this level, the focus is on the frictionless transmission of datum streams. Where the organization is large enough to operate additional nodal installations (according to its functional organizational structure), stage 3 is repeated for each one until the *plant* level is reached.

■ Stage 4: The roll-out team backtracks to stages 1, 2, or 3 to integrate, as the case dictates, the *datum streams* that are external to the organization. But caution is advised: before connecting to the external world, make sure that the safety and security controls are in place, operational, and monitored in real time. Otherwise, breaches from hacks and attacks could occur unbeknownst to anyone.

The reader may wonder why stage 4 is carried out so late in the game, instead of incorporating the external *datum streams* wherever they appear in stages 1 through 3. It is indeed possible to do just that, if the inner workings of a nodal system or installation require them. The primary concern here lies with the security of these transactions. Connecting to the outside world will always entail the risk of exposure of the organization to hacking, denial of service, and penetration attacks. If the digital construct's security controls are not up and running, you are leaving ajar an outside door that should be locked.

# The Digital Construct

## *The Organization's Holobiome Landscape*

The digital construct embodies the physical, algorithmic, and procedural features of the transformation. It is the foundation of the holobiome within which the organization will soon operate. It comprises the assets (hardware and software, together called "construct assets"), the interfaces (web, cloud, IoT platforms, holobiome networks), and the plumbing connecting everything together. The construct assets and the interfaces will, per force, be unique to the organization's design and buying decisions and administered, for the most part, by the usual IT group. On the hardware side, the prominent difference between the past and the pending future will be the gargantuan increase in data volumes. This new operating condition requires a thorough rethink of storage strategies away from the age-old central database setup; cloud solutions are bound to play a role in one shape of another. Data lakes, edge storage, and multiplatform data architectures render centralization impossible. Access will be solved by virtual interface strategies vouchsafed by new data-centric software applications (such as *Hadoop* and *Spark*). The plethora of constraints, risks, and threats pervading the issue of access also requires a thorough rethink in light of the inevitable heavy hand of regulators and legislators regarding datum privacy, ownership, and protection. Staying on the right side of the law, for the wild west days of unfettered data monetization are coming to a rapid end (thankfully), is taken up by a governance framework (see below). Such a framework MUST be in place before the pilot transformation is rolled out. Finally, the mechanics, mechanisms, and operability of the variegated assets, including mobile gadgets, IoT-enabled systems, and yet-to-be-invented new paraphernalia, fall under the aegis of data management (discussed below). Keeping all this equipment and these apps, algorithms, and licenses up to date becomes a far more challenging task than rolling out updates the old-fashioned way. As for the sophisticated beasts from the realms of AI and DLT, the primary hurdle will be human skillsets (discussed in the next chapter).

## The Prime Hardware Challenges

The bits and pieces that end up in the digital construct will be never be homogeneous, especially if some kind of IoT, even in the most primitive form, is in play. Fortunately, there are enough standards on the market to guarantee *interoperability* between them.* When edge computing is added to the mix, the issue of heat rejection and chip cooling must be embedded into

---

* The expression "interoperability" is a feature of a system (be it physical, algorithmic, or both) that enables it to connect to other systems (physically, controllably, and logically) such that their individual functions will operate without restriction or degradation when they are dependent on external functions for their inputs/outputs. "Plug-and-play" is the poster child of interoperability.

the design constraints from the get-go. But that too is sufficiently addressed by marketplace solutions, so that the organization should not have to resort to unique and costly custom designs. Information storage remains the organization's immediate constraint. The organization will be faced with multiple locations where data are stored across centralized databases, device memories, cloud-based servers, etc. Such dispersion will explode the intensity of the burden of administering operations 24/7. The maintenance strategy must be written explicitly with this dispersion front and center. As for the new hardware deployed throughout the organization, the logistics of maintenance, support, and repairs must be expanded well beyond the typical backroom warehouse ran by the IT department—especially if devices operate in transit (phones, sensors, GPS trackers, etc.).

## *The Prime Software Challenges*

Two software obstacles confront the aspiring digital transformer. The first is the proliferation of datum types, structures, and storage, summed up by the expression "multiplatform data architectures (MDA)". The second is data intelligence subtending governance. We address each one in turn.

The Symbiocene age brought an end to the age-old paradigm of the single data storage platform. Today's digital paradigm carries within it an extreme diversity of datum structures, operational functions, and business analytics algorithms. It is simply not possible to manage them all under a singular datum management environment. Instead, organizations must learn to unify the entirety of their siloed data *without first aggregating them.* The diverse data architecture portfolio is the new norm, with the organization's data intensively distributed among several storage designs (central databases, server farms, cloud locations, filing systems, etc.). Diversity becomes a strategic asset *provided that it can be corralled by a large-scale, unified cross-platform architecture.*\* Without it, the portfolio will rapidly devolve into a miasma of chaotic and unmanageable landscape of data.

---

\* The expression "data architecture" is generally understood to mean an all-encompassing orchestration of the datum models (format, nomenclature, file extensions, relationships, etc.), governance policies and management rules, standards and conventions, lifecycle management, usages and treatments, and user permissions that are to be applied ubiquitously to the generation, collection, and deployment of numeric data. The word "architecture" in this context pertains to the relationships that may or can exist between data and datum sets or, when speaking of "systems architecture", to the relationships between interfacing systems. It is important to remember that "data models" are local while "data architecture" is global.

The unified MDA seeks to eliminate the heterogenous ensemble of siloed platforms. It is a collection of relational and analytics databases, centralized file systems, and cloud storage services, integrated via a unified data architecture characterized by data integration infrastructures, metadata and semantics approaches,* data governance and visualization, and shared/homogenous data models. The unified MDA is particularly important to cloud environments that would otherwise be plagued by high latency and inelasticity of demand. Once again, the marketplace is ahead of the pack when it comes to scalable computing solutions for datum storage and transactions distributed across hardware clusters. The open-source application *Hadoop*, by the Apache Software Foundation (see https://hadooop.apache.org), is the dominant framework at the moment. Hadoop is widely used as the underlying foundation for orchestrating an organization's essential data functions (storage, filing, accessing, etc.). The most important aspect of Hadoop is its distributed file system, which sits above the file systems employed by computers and servers. This feature enables a user to access Hadoop from any system networked within the digital construct. Hadoop also comes with a comprehensive suite of analytical modules, which can be complemented or superseded by more specialized (and faster solutions) from the likes of *Spark* and *Solr* (also by Apache), and Azure (by Microsoft) to name but three. Predictably, implementing any of these solutions is not a task that can or should be improvised on the fly by the organization. Nor should the design of the data architecture. The field is a highly specialized one that demands a level of expertise found only with seasoned big data consultancies.

The second challenge deals with the more mysterious issue of governance (discussed in the next article). *Data intelligence* refers to the means needed by the governance framework to answer the W5H questions related to compliance:

■ Where are the data and what are their origins?
■ What does the data mean, imply, or signify?
■ Who has access to the data, to what extent?
■ When were the data last validated and verified?

---

* Wikipedia speaks of "semantic data models" as conceptual models of data based on semantic information (meaning that the meaning of the model's instances is defined). Source: Wikipedia. Semantic data model. Available at: https://en.wikipedia.org/wiki/Semantic_data_model. Last modified 30th March 2020. The model operates as "an abstraction that defines how the stored symbols (the instance data) relate to the real world". The model is capable of expressing the information needed by parties to transact data on the same basis of interpretation.

- Why do they exist in the first place (are they really needed)?
- How can data be utilized? And what are their inherent/emerging relationships?

The gist of these questions shapes the nature of the software applications employed to answer them. *Data intelligence software*[*] is essentially policing, in real time, the digital assets of the organization and issues warnings to management when suspicious or out-of-the-norm transactions are detected. Usually, there will be more than one application operated concomitantly for the purposes of data discovery, cataloging, profiling, master definitions, ancestry, effectivity, and stewardship. The aim of this collection of applications is to quantity the information needed by management to govern the organization's digital assets.

> Data intelligence software helps governance officers to enforce the rules of engagement by everyone in the organization. They are not there to optimize processes or monetize the digital assets.

IDC, a consultancy, recommends four industry frameworks for helping organizations deal with data intelligence solutions:

- DAMA (Data Asset Management Association), especially its book of knowledge
- CMMI Data Management Maturity
- EDM Council Data Capability Assessment Model (DCAM)
- MIKE 2.0

## The Construct Manifest

The administration of the digital construct requires a thorough knowledge of what is installed where and doing what for whom—presumably, a self-evident fact that needs no furtherance. Nevertheless, the path to quantifying this knowledge is not readily evident. It begins with an inventory of all the bits and pieces making up the digital construct: things like part numbers, description, supplier names, costs, serial numbers, physical location, specifications, purchasing history, etc. The inventory is necessary but not sufficient. We need

---

[*] See IDC PlanScape: *Data Intelligence Software for Data Governance*, Report # US41714817, August 2018. Available at: https://www.idc.com/getdoc.jsp?containerId=US41714817.

a construct *manifest*, of which the inventory is but a subset. All these bits and pieces must also be documented in terms of their functions within the digital construct, along with their UTP characteristics (see Figure 3.4). It begins with the assignment of a unique digital construct identifier derived from the classification system (see Chapters 3 and 4). It is followed with the details of the physical, algorithmic, and operating specifications, the technical datasheets and associated documentation, the maintenance and troubleshooting documentation, the reliability and maintainability requirements, the reliability track record, the interchangeability and interoperability profiles, the maintenance/ update cycles, the user and administrator instructions, the user and administrator training syllabus, the firmware version tracking, the *model calls* directly involved, and the organizational accountability assignments. The reader will appreciate by now the extent of the *construct manifest*; it is not something that can be addressed as an after-thought or improvised at the last minute. It is in fact a comprehensive process that aggregates all this information over time, as the transformation team progresses across phases I through IV.

The last piece of the *manifest* deals with the testing, validation, and verification of the bits and pieces making up the *digital construct*. Physical pieces will be bought, assembled, and tested within their nodal systems or installations to make sure that they work symbiotically. Revisions to the datasheets, selection criteria, and pertinent specifications may be needed if problems are discovered. Software pieces may also need to be checked for operability, version compatibility, user experience, and performance expectations. Interoperability, ease of installation/maintenance, and vendor support can also be checked. Advanced applications such as data intelligence and artificial intelligence applications may require individual testing and validation programs, separate from digital construct interfaces, to decide which ones to buy or avoid. The results of these actions must be documented in the *construct manifest* to provide a decision history on why something was selected instead of another.

# Corporate Digital Governance

## *One Framework to Rule Them All*

Governance is about making sure that the organization does not run afoul of the law or regulators. Governance erects the boundaries of the organizational framework through which the assets are exploited profitably. And just be sure that the reader's attention is focused adequately, consider the fines carried by the General Data Protection Regulation (GDPR), rolled out across Europe in 2018. An organization that is found to be non-compliant with any part of the GDPR can be fined up to 4% of *its annual revenues*. The strictness and heavy handedness of the GDPR arose, in part, in response to the grievous breaches of personal data uses by the likes of Facebook and Google in the years leading up to the GDPR's release. And make no mistake about where compliance is going, worldwide: in the West, it is tilting bluntly on the side of individuals and other data generators, at the expense of firms that collect the data.

Governance, in other words, is unavoidable: *thou shall govern your digital operations*. Policing is an integral part of governance, regardless of the harshness of the word. Governance exists first and foremost to obviate the risks to one's liability, regulatory non-compliance, and legal turmoil.

> Governance is not a technology-based solution. In fact, it exists independently of an organization's technological footings. Governance is a corporate management concern resolved as an organizational framework aided by technology.

## Governance Policy

Governance is not like management fads (diversity, inclusivity, sustainability, de-carbonization, and the likes), which are often initiated hastily by corporate boards in response to the communal value-signaling du jour.* The onus for governance sits squarely with the CEO, on par with financial accountability. A single instance of a very public disclosure of a data breach should suffice to focus the board's mind on the importance of it.

The principles of governance are formulated into a corporate policy implemented uniformly across the organization, under the authority of the CEO. Once it is in place, the policy is administered by the Chief Information Officer or equivalent (more on this in Chapter 8). The form and extents of the policy is articulated as a function of the organization's size, digital operations, and business needs. It must be written concisely, with as few mandatory clauses as possible so as to leave room for the organization to operate dynamically within in, and in keeping with advances occurring on the

---

* A board will suddenly scramble to assemble a response team, order the release of grand public pronouncements about their newly found virtue signaling, then carry out these programs *sotto voce*, until such time when media histrionics have abated. At which point, the initiative is quietly wrapped up without fanfare or lasting organizational changes.

outside. One effective way to develop the policy is to divide its contents in four sections:

- Binary strategy section, speaking about the overall objectives of the organization, the types of digital capabilities it must have, and the connection between those capabilities and the long-term commercial/financial objectives of the organization. The strategy should ideally refer back (through an annex or appendix) to the *model calls* (developed in Chapter 3)
- Compliance section, comprising an explicit declaration of unconditional compliance to whatever laws, regulations, and national codes that are pertinent to the organization's operational footprint. The section must establish (via an annex or appendix) a detailed hierarchy for direct accountability regarding compliance assurance and verification, non-compliance discovery, and breach remediations. Finally, the section shall include a position statement with respect to data ownership and the mechanics of enforcement and breach discovery.
- Execution section, spelling out who's who in the digital zoo (along with the obligatory organizational chart), and presenting a list of guidelines to shape the behavior of the staff in their dealings with data. The section must include a *directrix assignment table* (pursuant to the discussion from Chapter 3) with respect to the functions and tasks associated with the overall management of the governance portfolio.
- Standards section, spelling out or calling to be developed the standards, procedures, and processes to be rolled out across the organization to operate the digital construct. The section must specify the *directrix* assignments for each item. The section must also spell out, directly or via a reference, the lifecycle management process (creation, publication, updates, senescence) for all standard items.

Keep in mind that the *governance policy* document is not a vision or mission statement. It is not meant to be inspirational or aspirational. It is, above all else, prescriptive and coercive.

## Pilot Implementation

### *The Right and Wrong Way to Proceed*

The governance schema must be in place and in force before the pilot transformation is ready to come online, operationally. Proceeding in reverse will prove wasteful and costly.

Everything that has been written up to this point served to define, constrain, and specify the details of the digital construct for the organization. Some of the details will be sorted out from actual tests and dry runs (see the previous article) but nothing has really been rolled out operationally. Certainly, the essential principles, players, and rules of the governance framework will have been nailed down. *Do not proceed in reverse by rolling things out first then develop the digital governance.* Remember that governance very much acts as inputs to the design of the construct. Remember also not to allow your *vision* to become a binary organization to belittle what it will take to get there. The allure of a binary future is enthralling to the visionary business leader and the enthusiastic entrepreneur; but that future will not be achieved in one fell swoop, through a one-time transformation initiative spearheaded by the board of directors. The level of effort to get there scales exponentially with the number of information nodes involved. The scale is also a progenitor of the complexity risks underlying the transformation journey: the broader the scale, the greater the risks, and even more so when the organization is new to the game. Prudence suggests an incremental approach to rolling out the digital transformation, rather than attack the entire organization from the outset. Hence the pilot transformation to limit costly risks and allow the organization to "learn to walk" the transformation process before going all out across the organization.

> The riskiest approach for both novice and experienced organizations is to undertake the transformation for the entire operations from the outset, with an expectation of implementing the strategy concomitantly across all functions.

## A WORD FROM YOUR AGILE SPONSOR

What about the "agile methodology", prevalent in the software development world? Can it not be adapted for the purpose of implementing the digital transformation? After all, the method has been the dominant school of thought over the past several years (i.e., code, fail fast, change, repeat). Its appeal is understandably appealing: the method advocates the concomitant evolution of the requirements of software along with the solutions to their embodiment through the equally concomitant works of cross-functional development teams, their customers, and the intended end users. By contraposition, the incremental and sequential approach described in this book is explicitly linear and reminiscent of the waterfall model for software development. It is slower and more structured than the agile approach. The latter is highly risky to the newcomer, especially one not belonging to the software industry. When dealing with organizational transformations, failing fast, changing, and repeating is a recipe for disaster. Going about things incrementally is inherently safer and far less risky than going agile.

## Where to Begin

Which functional area of the organization should be chosen for the pilot transformation? Remember that this first time out will be nothing but complicated, inefficient, frustrating, slow, and perhaps costlier and longer than planned. These growing pains are unavoidable. It is the nature of complex systems anchored to software to be fraught with bugs and hurdles during installation and commissioning. The decision must balance the need to validate the digital investment without hampering the commercial viability of the organization. So, which one to choose?

For this purpose, we bring back our fictitious firm IFL from Chapter 4. We assume that IFL sells its wares in the Americas and Africa through a network of satellite offices manned by independent operators. Assume that IFL comprises the usual departmental suspects—sales, production, engineering, warehousing, administration, finance, and senior management. Each department is likewise divided into functional groups, each responsible for a specific mandate. The best candidate will usually be the sales department. Sales are the department that acts as the information nexus for the organization. It is equally influenced by exogenous and endogenous data (see CLASS in Table 3.1); it is at the receiving of virtually all prime and derivative data (see the ORDER in Table 3.1) produced internally and collected externally; it impacts directly the cost and revenue information (see GENUS in Table 3.1); and it transacts all information species with the organization's customers. Last but least, the implementation is the least likely to affect overall daily operations. Since the sales department is mainly a *receiver* of information inputs (in the UTP sense) relative to its output production, it is the least likely to disrupt the information transactions of the other departments.

For very small organizations (typically less than 25 or so employees), it is conceivably feasible to entertain transforming all operations at once, albeit always under the shadow of risk. One should not fall prey to the perception that small is easy; even a 20-person, office-less organization can hide unsuspecting levels of complexity, and therefore risks, that could derail an all-encompassing initiative. In the end, choosing the scale at which the digital transformation is enacted boils down to the firm's ability to proactively manage the complexity of the transformation journey.

---

### THE INVESTMENT-CENTRIC APPROACH TO ROLLING THINGS OUT

While the details of both pilot and full rollouts will be unique to each organization, the sequencing of the work will generally follow the same road map (and violently exclude any "agile" ideas). The reader interested in the mechanics and mechanism involved with this road is invited to consult Chapters 3 and 4 of *Investment-Centric Innovation Project Management*. The approach described in this chapter implicitly abides by the investment-centric method.

---

## What about the Digital Construct?

The pilot transformation will require a partial implementation of the *digital construct*. Which parts exactly will be dictated by the input and output streams of the sales group. The underlying plumbing of the construct will already be in place anyway, assuming that the organization was hitherto running some kind of IT network. The pieces required by the group's endogenous *model calls* will be the ones to be installed from the outset. What is more important, however, is the training of the sales staff (and the IT group, in its supporting role) to operate the various pieces of the construct. Such training should be conducted before the pieces become operational, but as close to that start-up date as possible. While the pieces are put through their paces, by said personnel, system matter experts (already trained super users or consultants or vendor representatives or all) must be on hand to assist the neophyte users. The transformation team remains in control of the pilot implementation throughout the endeavor.

## The Learnings

The pilot transformation serves one more purpose beyond that of implementing the digital transformation at a small scale. It is to teach the organization what works and doesn't and what needs to change to improve the process next time around. Each problem, each bug, each setback, and each user action failure is an opportunity to pause and reflect on why it appeared, how to solve it, and how to avoid it in the future. Each opportunity is a lesson, big or small, that should be captured in writing and cataloged by the transformation team in a dedicated pilot lesson binder (or whatever else the reader wishes to call it). Lessons should be cataloged according to who encountered them the first time around, an approach that lends itself to producing a natural taxonomy defined by skillset or function. The user category will be especially revealing with respect to the completeness of the training, the unforeseen problems with human-machine-software interactions, and the adequacy of the graphic user interfaces. All these observations and insights should find their way into the appropriate documentation (user manual, training manuals, operating manuals, etc.) developed internally or bought from vendors. Another category of importance will surround the issue of tricks and workarounds for the installation of devices and software, and the deficiencies noted in the product literatures supplied by vendors. Finally, the vendor support category will prove priceless in capturing the insights of how to best seek and obtain technical support from these vendors, even if the conclusion is that a vendor is near useless in that role (in which case one may wish to change things up before the full roll-out goes forward).

## Post-Mortem Assessment

The closing act of the implementation is an assessment of the overall outcomes of the pilot transformation. *This assessment is sine qua none.* Its aim is to take a step back from the nitty gritty details and perform an honest yet critical review of what worked, what did not, and why, and set out the supplemental work that will be required before the full roll-out is undertaken. The review is not about finger pointing, but about value validation, as in "is something worth continuing to be deployed moving forward?" For, there may be several elements of the digital construct that could not be made to work as expected or made to work with great difficulty without corresponding benefits. Some of the theoretical benefits/gains may turn out to be insignificant or simply too onerous for the value obtained. Or, it could be that the current state of the organization's digital potential is simply too primitive to be able to transition to the intended digital prowess in one fell swoop. For instance, an internal IoT framework may have sounded awesome as a solution to a *model call* but requires a level of infrastructure revamp that costs too much at this particular point in time.

The review is completed via a debriefing meeting between the leadership of the transformation team and the organization's senior management team. The agenda should be divided into two parts, Assessment and Decision. The first part addresses six questions:

1. What was shown to work, is ready, and recommended for full roll-out.
2. What was shown to work but is not ready for full roll-out because of underlying IT infrastructure deficiencies (why, what is required to make it ready, at what cost and time, and when can it be rolled out).
3. What was shown to work but with difficulty and is worth pursuing *at a later time beyond the completed full roll-out.*
4. What did not work but could be worth pursuing (why, at what cost).
5. What was shown to work but is deemed too onerous for the benefits observed.
6. What did not work and is not worth pursuing.

The second part seeks to obtain the go-head from management to proceed with the follow-on work. For every scenario listed below, the transformation team needs to present the work plan, the timelines, the costs, and the resources required to complete.

1. For question 1, the follow-on work is the full implementation roll-out across the organization.
2. For question 2, the work is a second (or third or fourth) pilot trial.
3. For questions 3 and 4, there is no work for now. Wait until full roll-out is done, then circle back to the plan and cost and time to get it done.
4. For questions 5 and 6, the work consists in going back to the formulation of the impacted *model calls* and explore different formulations and/ or solutions (either case will require a pilot trial later).

# Full Implementation

## *The Plan*

The full roll-out of the digital transformation is a significant endeavor that affects the entire organization. This is step 4, summarized earlier in the introduction, and is carried out in four stages of increasing nodal complexity, even when the work proceeds sequentially, one functional group after another. The initiative will take shape as a major capital project for the organization, which means extensive *planning, budgeting, scoping,* and *scheduling.* Improvisation (ad hoc or agile-driven) will be the death knell of the enterprise. Once again, incrementalism governs.

The scope of a full implementation is beyond the space available in this book. On the other hand, we can discuss salient features aspects that will be common to most of them.

**Governance**. The plan must discuss how the governance framework will be deployed across the organization. Organization charts are required for everyone to see who's who in the digital zoo. The procedures, standards, templates, and other like-minded documentation must be published uniformly, with read access granted to those who will be impacted by them. Individual, management, and group training sessions will be required as well, to create a baseline understanding among everyone about how governance will be managed, enforced, and dispensed.

**Digital construct**. The construct was only partially constructed during the pilot phase. The remaining parts now need to be constructed, installed, and tested. The scale of this work may justify the execution of the work under a separate capital project of its own. The initiation of the work should be preceded by an announcement to the entire organization about the nature of the work, the reasons for it, and the key milestone dates of completion. The announcement should list what will be added, removed, improved, and inaugurated (where new capabilities are concerned). The direct accountability assignments for the management and administration of the construct should be published from the outset. Like governance, all documentation must be published uniformly, with personnel training orchestrated according to the implementation schedule. The master schedule must be published widely and updated regularly with progress metrics and newly activated capabilities.

**Schedules**. It behooves the transformation team to engage everyone in the organization by publishing regular updates on the progress of the works. An integral part of these updates includes two schedules, one for transformation milestone schedule and one for personnel training. The latter is the more important of the two because it is the direct link between people and their involvement with the new digital framework.

## The Missing Keystone

The natural outcome of the work prescribed in this chapter is a digitally transformed organization. From this point forward, it will operate as a genuine binary framework ready to start delivering on the commercial objectives posited from the outset as justification for the journey. Together, Chapters 4 through 7 provide the reader with a comprehensive method for planning and executing the digital transformation with the implementation road map detailed in this chapter. The reader will still require the assistance of technology specialists to translate a *model call* into a physical solution (Chapters 5 and 6) and deploy the solutions into their operational settings (this chapter). Always keep your consultants on a short leash within a well-defined mandate. They are there to assist you, *not to lead you where they think you ought to go.* The same caution goes for engineers, who, if left to their own devices, will gleefully design phantasmagorical concepts like toddlers playing in a muddy puddle.

---

**THE CONSULTANT'S MANTRA**

The all-time best caution about hiring consultants was postulated sardonically by the brilliant company *Despair, Inc* (www.despair.com), on its demotivational poster *Consulting*: "If you're not part of the solution, there's good money to be made by prolonging the problem".

---

The other people involved in the transformation were generically lumped under the *transformation team* umbrella. Which does them quite a disservice since those people make-up the keystone of the entire transformation edifice. Despair not, however: the next chapter is dedicated to them.

# Bibliography

Keays, Steven J. *Investment-Centric Project Management: Advanced Strategies for Developing and Executing Successful Capital Projects.* J. Ross Publishing, Plantation, FL, 2017, 419 pages.

Keays, Steven J. *Investment-Centric Innovation Project Management: Winning the New Product Development Game.* J. Ross Publishing, Plantation, FL, 2018, 309 pages.

# THE BINARY ORGANIZATION

# Chapter 8

## People Digital

*All the robots, AI, neural networks, and whizbang technologies won't amount to a hill of beans in this crazy world that's banking on cutting humans out.*[*]

It's a brave new employment world out there. Here, a synopsis of the new jobs, the new skillsets and the new management paradigms ushered in by the Symbiocene age.

## The New Labor Landscape

### The Binary Worker

The most important success factor in any digital management transformation is the human element, bar none. All the technologies, all the new processes, all the new capabilities will amount to naught if the organization fails to grasp the centrality of people to the endeavor. We just acknowledge and recognize the fear of dislocation and unemployment that pervades among working people everywhere. The overwhelming message conveyed loud and clear, far and wide, by innumerable news stories is the inevitability of digitalization in the workplace, to the *detriment* of workers caught at the wrong end of digital incursions. The perception, rightly or wrongly, has hitherto

---

[*] A modern twist inspired by a famous line from the eternally famous movie *Casablanca* (released in 1942).

197

evolved into a nasty, brutish winners-take-all zero-sum game titled in favor of the organizations willing to embrace digitalization and cull their employment rolls. Most unfortunately, what's heralded as fantastic business opportunities with each new digital application is inevitably couched in warnings of vanishing jobs, discarded skillsets, and perennial unemployment for large swathes of working populations.

There is no denying the inevitability of technological incursions into all manners of human activities. However, the dystopian notion that the digital hegemon will conquer all and leave millions of job casualties in its tracks is simply wrong. For societies (at least in Western democracies) will revolt against such draconian consequences through government policies.* There will be job losses, as is always the case with capitalism's creative destructions. Repetitive tasks, on the desk or the shop floor, are already on their way out. Monotonous work like driving long-haul trucks will be taken up by computers (but the last miles will remain unabashedly in human drivers). "Discovery" and case research will be left to AI applications while lawyers focus on the strategy to make their cases. The list goes on an on. The nature of work has changed already and will continue to do so in ways that are unfathomable. What is equally unfathomable is the idea that a business can operate on digitalization alone, with humans running the C-suite and reaping all profits.

Perennial success in the Symbiocene age demands a symbiosis between the digital construct and employees (taken, in our present context, to encompass payroll, contractor, and consultant staff). The most profitable operators are those who look upon their digital construct as powerful (but tamed) *enablers* of employee performance. They recognize the benefits of delegating to machines and algorithms the functions for which humans simply cannot accomplish competitively (i.e., in competition with the machines and algorithms). It makes no sense, economic or otherwise, to employ someone in the scanning and recording of receipts to prepare an expense report, when a piece of software loaded on a smartphone can do it via a simple picture taken by the person incurring the expense. Nor does it make any

---

* The simplest and most effective counter measure that a government can take to protect its population against seditious capitalism is to design a corporate tax framework that is defined solely and exclusively around the notion of maximum human employment remunerated with maximum contributions from a firm's gross profits. Eliminate all deductions other than direct employment costs (minus executive compensation, which distort the remuneration statistics). In such a scenario, a firm that maximizes digitalization at the expense of employment will see its tax extorsion (for such is the correct term for all forms of taxation) kill its bottom line.

sense to employ a costly AI application in a customer service function when a caller's issues need resolving at the human level. But imagine the efficacy of the process if the AI is subservient to the customer service representative, now equipped with the ability to search and analyze in real time possible solutions garnered from a historical database.

> The working symbiosis of people and digital is a whole that exceeds the sum of its parts.

## The Mechanics of Symbiosis

Let us state at this point an axiom that should be already self-evident:

> To a binary firm,* people matter more than the digital construct, the more so as digital complexity increases.

This is the most important insight of this chapter, if not the entire book. The importance of people grows in magnitude as a function of the complexity of the digital construct. The greater the complexity, the more important people are to the long-term health of the firm. Three questions confront us now:

1. What does "symbiosis" look like?
2. What does "working symbiosis" mean?
3. How does one achieve such symbiosis?

Symbiosis is visible when the worker utilizes the features of the digital construct that are available to him instinctively and naturally without giving it a second thought. In fact, the reader is already acquainted with the notion through one's smartphone or email application or laptop. In each instance, the technologies involved are utilized as a natural extension of the user. The user does not wonder about what else to use, for it is implicit to the user's *modus operandi*.

The second question is also answered by these three examples. The symbiosis is "working", in the sense of being efficient, productive, and

---

\* From this point forward, the text will utilize the term "firm" instead of "organization", as has been the case so far, to focus the reader's mind to the monetary bottom line ramifications of the transformation. Even governments, NGOs, and charities must operate within their budgets, regardless of their funding source(s).

specifically tailored to the end goal sought. The interactions between the user and the technologies are frictionless, the outcomes come out as expected, and in complete agreement with the anticipated time and manipulations required to achieve them. By contrast, a symbiosis that is not frictionless, or produces unexpected outcomes, and requires more time and manipulations than imagined, or all three, is not working. Daily life is replete with example of non-working symbiosis: the clunky website that is anything but instinctive; the help line that puts the caller on hold for 40 minutes; the new computer program that crashes without explanation; a remote control with too many buttons and functions to make sense.

The notion of a "working symbiosis" takes on an entirely different importance when dealing with systems of greater complexity, such as casually querying an ERP system or attempting to create a graphical presentation of the results of a deep learning algorithm. If the user cannot get to the desired outcomes efficiently, or is required to go through cumbersome workarounds and overly complicated menus, or requires assistance from a dedicated super user/expert, the symbiosis is clearly not working.

> It is not enough for a user to be able to interact with a technology by finding ways to make it work somehow. The usefulness of the technology will be inversely proportional to the amount of friction experienced by the user. When it is too hard or time consuming to use, it simply won't be used, unless forcibly compelled. Friction is the ultimate technology killer.

The answer to the third question involves five steps:

1. Design (or select) a GUI that is intuitive, obvious, procedurally parsimonious, queried via natural language, and opaque to the user in terms of the underlying technology. There should not be the need for extensive training, for learning a unique machine language to interact with the technology, or for learning the inner workings of the algorithms working in the background, where the real computing takes place. If such a GUI is not feasible, the technology will only be available to trained specialists—which is acceptable in specific circumstances but not as a general-purpose technology.
2. Assemble and train a cadre of superusers on the technologies to be deployed both locally and widely within the firm. For the widely used

ones, disperse the superusers among the staff to act as future go-to resources.

3. Train the users, in four parts. Part 1 occurs before the user employs the technology in an operational setting, when training is focused on the basic interactions between human and technology. Part 2 places the user in an operational setting, *with real-time access to trainers/experts in case problems arise.* Part 3 returns the user to the classroom for training in advanced features. Part 4 repeats the assisted experience of Part 2.

4. Establish a series of user forums two weeks after initiating Part 2 of the training and invite users to attend them to exchange their experiences, discuss problems and workarounds, and propagate learned lessons quickly. Team up the people who are experiencing difficulties with people who are not in an informal mentoring relationship.

5. Three months after the start of training Part 2, and after the start of Part 4, re-mobilize the transformation team to conduct audits of the technologies in their operational settings. Are they meeting the expectations? Are they proficient in delivering the *model calls?* Have they achieved "working symbiosis" for the majority of the workers? If not, why? Allow the findings to dictate the next course of action, which could be, in extremis, termination of the technology.

## *The Importance of Being Gray-est*

It is often the case that the younger employees have the easiest time with new technologies. Nurture this benefit and put it to work by twinning them with seasoned employees. At the same time, tap the wealth of expertise buried inside the brains of these seasoned employees to verify that the new technologies, and the people using them, are really improving the operations and the processes. NEVER trust blindly the purported suitability of a technology for the intended operational purpose. Only the gray owls in your organization will know if what comes out is worth what was put in.

> The worst possible thing to do is to shove the gray owls to the curb and hire squadrons of (cheaper) younglings presumed to be more adept at new technologies. No software, no matter how sophisticated it is, understands what works, what does not, and why inside your firm. That understanding lies quintessentially with the seasoned employees.

# Managing the Binary Firm

## *Weaving the Proper Fabric*

We come now to the issue of managing the digitally transformed firm, which requires a dedicated management framework at the corporate level. The framework does not need to be bulky but must fall under the purview of an executive who is plugged into the C-Suite. Digital management is different than its organizational brethren because of the unique DNA carried by the digital construct. This DNA renders everyone that uses it visible and traceable. In other words, people in a binary firm must contend with the potential for being surreptitiously monitored in real time by faceless eyes. This distinction must be addressed from the outset to placate any attempt by prying bytes to reign over employees. It behooves the firm to weave into its management philosophy the following covenants (without which the place will reek of suspicion, distrust, and excessive turnover).

1. *People first.* Humans precede technology on the list of priorities. This precedence increases in importance with increasing levels of technological symbiosis. The minute people are subjugated to the ascendency of technology, they will begin to resent or revolt or worse, sabotage, the digital construct. In spite of the extreme prowess of technology, its benefits can only truly accrue to the firm through the agency of the people who make it all work together.
2. *People are people first then assets.* Having at one's command the ability to monitor by stealth every move of every person plugged into the digital construct is not justification to act upon it. The rights to privacy, against wanton searches, against arbitrary intrusions by an employer do not magically vanish at the threshold of the front door, whether as an employee or a visitor. The appalling example of China's dystopian social credit system exemplifies everything that is intolerable in the workplaces of free people. The drive for data privacy, heralded by Europe's GDPR, will inevitably expand into the realm of the workplace as well. Look at it this way: how long would a firm continue to retain its people if it was found to be secretly spying on them as they go about their daily office business?
3. *People are assets, not resources.* The old, vapid cliché that people are a firm's most important asset is too often brandied about fecklessly and quickly castrated when cost cutting is in play. But in a binary firm, it is an inalienable truth: people become the bulwark by which a firm protects itself from the vagaries of technology.

4. *People are profit—not cost—centers.* This the corollary to the previous covenant. The full potential of the binary firm percolates through people's mastery of its sinews. If payroll is nothing but a line item on the expense side of the general ledger, the true measure of its investment value will walk out the door when you cull it to protect the share price or the executive incentivization plan. A binary firm stripped of its human capital is a dead firm walking.

5. *People are not data.* Extreme visibility exposes everyone to a potentially excoriating light. Do not delude yourself into thinking that you can extract better overall performance by pitting everyone against everyone else on the basis of statistics derived from your own big data. Forget the bell curve (illustrated in Figure 8.1) as an excuse for indiscriminately pruning out the low sigma crowd. Work instead on both moving the average to the right (via training, continuous development, and process tweaking/optimization) and tightening the standard deviation (by better team distribution, intrateam mentoring, and team-based valuations).

6. *Technology is a part of, rather than be, the solution.* This is the corollary to the previous covenant. Technology can be many things, do many things, and exceed many humans in myriad ways; but it can never become an end unto itself at the exclusion of the human element. Even where only machines dare (or can) dwell, nothing can ever be so automated at the total exclusion of human interactions, owing to the insurmountable "exception handling" caveat, which separates programming from sentience.

## A BRIEF GAUSSIAN HISTORY

The normal distribution, affectionately called the bell curve because of its eponymous form, is a widely used statistical probability distribution in the hard and soft sciences. The curve presents at a glance the average "$\mu$" of a group of datum points and how the points cluster around it (measured by the standard deviation "$\sigma$"). In group without any deviation ($\sigma=0$), each element of the group has the same measure as the average and the curve collapses with a vertical line at the average. Otherwise, it takes on the familiar bell shape, where 68% of the datum points lie within one "$\sigma$" and 95% within two "$\sigma$". The distribution was first studied by the famed mathematician and physicist Johann Carl Friedrich Gauss (1777–1855), after whom the distribution is named. Gauss was to mathematical physics what Feynman was to quantum mechanics: a giant among intellectual giants.

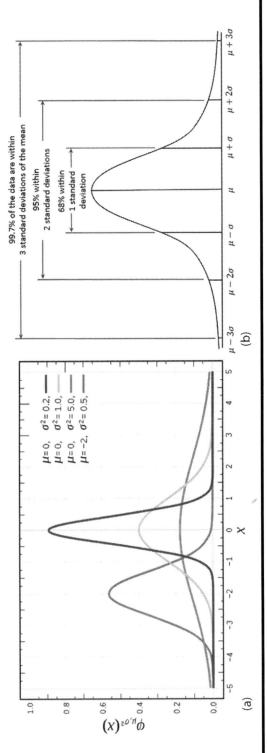

**Figure 8.1** The Gaussian portrait. The generic form is shown on the left. The insights are shown on the right.

## Make the Covenants Make You Money

The joint ramification of the fifth and sixth covenants endows the management framework with the ability to quantify the "working symbiosis" of an employee with the digital construct, via two measures: *interface efficacy* (how fast a user operates the functions of a specific technology or software *without errors*) and *functional density* (how many features/functionalities of the technology is exploited by the user). For instance, nearly everyone will be familiar with Microsoft Excel and be able to generate datasets of various complexity and presentations. How fast a person can generate a graph from tabulated data gives a measure (in units of time) of that person's *interface efficacy*. Being able to create a VB program within Excel that can, at the click of a "button" inserted into the worksheet, assemble the data from external spreadsheet sources then create, on a separate worksheet, the corresponding graph (fully formatted automatically) represents the measure of the person's *functional density*. These measures can be compiled for a group, a department, or the whole firm; these measures will invariably lead to a Gaussian distribution.

The beauty of this outcome is that it instantly provides a wealth of insights to management in terms of opportunities to raise the overall "working symbiosis" of the underlying group. The competent manager will infer five immediate opportunities:

1. Who are the proficient users (beyond the $+1\sigma$ mark in Figure 8.1b).
2. Who needs help (between the $-1\sigma$ and $-2\sigma$ band).
3. Who struggles and needs serious re-training OR be let go (to the left of the $-2\sigma$ mark).
4. Who are the super users (beyond the $+2\sigma$ mark).
5. What additional training can be cobbled together from the super users' experience to enhance the performance of the bulk of the employees (inside the $-1\sigma$ to $+1\sigma$ band).

What might an incompetent manager conclude from the same outcome? Actions such as summary firings of the struggling individuals, followed by a mindless edict to the rest to raise their game to close the gap with the top performers, under the threat of more firings if the average does not improve within an arbitrarily set timeframe.

Guess which approach will have the best impact to the long-term bottom line of the firm?

## Managers, Leaders, and Leaderers

It has been emphasized time and again throughout this book that a digital management transformation alters an organization in a most fundamental way. One cannot achieve this transformation merely by tweaking the existing management philosophy by adding a digital construct silo to the corporate organizational chart. Carrying out the transformation requires genuine leadership, anchored at the human level. This is said in contraposition to vesting the onus of the transformation on the shoulders of existing managers. The distinction between a manager and a leader takes its cues from the concepts developed in Chapter 3 of *Investment-Centric Project Management*. In essence, the leader stands where he sits, while the manager sits where he stands.

The success of the transformation is propitiated by leadership, individual and organizational. Titles and positions do not a leader make. The CEO is, by nature, a corporate creation thus a manager by default, a leader by exception. Evidently, the top dude or dudete can certainly make a unilateral declaration that a digital transformation shall take place and that's that. People are not stupid: they'll go along with any plan out of a primitive sense of self-preservation. Whether or not they will actively wish and work for its success is an entirely different story. That is where leadership comes in. Leadership will instill a sense of purpose and mission among the troops to make the journey a success.

> The fundamental axiom of management: people are led and everything else is managed

Unfortunately, the business literature is blindingly enamored with the idea of leadership as the end-all-be-all answer to business success. Manager and leader are conflated indiscriminately into meaning the same thing, as if being a senior executive automatically and inevitably implied a human leadership character. As a result, some managers (who are perfectly competent in their functions) will assume themselves to be leaders as well and delude themselves into believing themselves able to lord it over their subordinates with unquestioned authority. When this dichotomy arises between the manager's self-sense of leadership and the contrary perception of subordinates, the result is an ugly hybrid archetype called "leaderer".* The clearest

---

* See Chapter 3 of *Investment-Centric Project Management.*

indication that a manager is operating as a leaderer is via a propensity to manage predominantly by numbers, which leads to making decisions based on data analysis bereft of the human context. Sadly, when leaderers are allowed to roam free, they can infect an organization like a tumor spreading across the rank and file, sometimes into the ground.

> **NUMBERS ALWAYS MATTER**
>
> The diss about "leaderers" does not imply that managing by the number and the bottom line is wrong. On the contrary, it is imperative that the financial health of the firm is continuously monitored by senior management.* In difficult times, there may be no other choice than to lay off employees as a short-term survival tactic (but never a strategy). Faced with such a decision, the leader will impose the pain from the top first and flow it progressively through the lower ranks until the situation is stabilized. Those at the bottom of the hierarchy will be the last to be affected. When good times return, the leader will reward the troops in reverse order, from bottom to top. With a leaderer, the start and end points of the pain flow and reward wave are reversed, such that the top always benefits first, the most.

# Management Framework

## *Dual Structure*

The dedicated management framework was introduced in Chapter 7 under the heading "corporate digital governance". The framework is divided into two realms of accountability: *holobiome oversight* and *construct management*. The former is hegemon over the entire organization, with a focus on compliance of everyone and every activity to the policy. The latter is concerned with the bits, pieces, mechanics, mechanisms, and machine-human interactions implicated in the daily doings of the digital construct.

The overall management of the framework needs to be delegated to an individual able to dedicate the time necessary for it. This could be the Chief Information Officer, the Chief Technology Officer, or suitable others at this level. This is an organizational position, as opposed to a functional one,

---

* The "leaderer" archetype is germane to this book but not discussed in Chapter 3 of *Investment-Centric Project Management*.

which must be assigned by name in accordance with direct accountability principles. The ultimate purpose of the position is threefold: enforce the policy uniformly, enable and empower the three members of the operational team (see next) to fulfill their respective mandates, and maintain clear communication interfaces with the executive team.

The operational team includes three roles reporting directly to the CIO, and shown in Figure 8.2:

■ The Chief Data Officer (CDO), directly accountable for the processes, procedures, and systems involved with data creation, storage, cataloging, lineage, data profiling, and master definitions (including data architecture, semantics, interoperability). Data, algorithms, and technologies are at the core of this role but are not the role's purpose; the purpose is to deliver the expectations of the *model calls* and vet their continued relevance to the strategic targets of the firm. The CDO bridges the chasm that often exists between the firm's decision-makers and its operational staff, busy on keeping the machine running. A potent CDO prioritises on customer service delivery, cognizance of existing and new market trends,

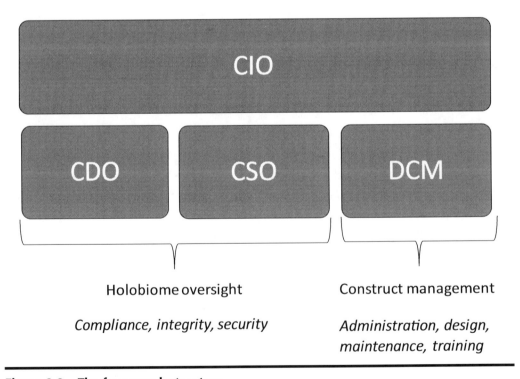

Figure 8.2 The framework structure.

continuous improvements of the digital construct, expanded revenue streams, and shareholders' perennial ROI. Data, omnipresent and never-ending, are the grounds where these priorities grow.

■ The Chief Security Officer (CSO), directly accountable for the integrity of the digital construct, including protection against external agents, prevention of internal sabotage, threat detection and mobilization, and hacking countermeasures. The purpose of this role is self-evident: protect the firm against cyber threats both from within and without. The role however goes beyond that of digital police and comprises three additional components: 1) bulwark defender (the firewall, digital moat, and denial team to combat external, active threats); 2) endogenous traffic monitoring, suspicious activities detection, and real-time user alerts; 3) continuous training of personnel on embedded security and counter-insurgency measures; 4) keep up with advances in the state-of-the-art defence mechanics and mechanisms; and 5) conduct recurring SWOT assessments of the digital construct's state of protection.

■ The Digital Construct Manager (DCM), directly accountable for the operation, maintenance, and administration of the digital construct infrastructure (hardware, software, training). The DCM's mandate is straightforward: keep the digital construct running smoothly. The mandate includes acquisition, installation, operation, maintenance, troubleshooting, lifecycle management, personnel training, and "working symbiosis" tracking of individuals and groups alike. The DCM is also accountable (in a *directrix* sense) for keeping up with advances in the marketplace of the construct's components as they arise over time. Finally, the position includes an advisory function to the CIO about the state of the digital construct, its efficacy and shortcomings, and about options for changing it in accordance with the changing state-of-the-art in the broader marketplace.

The matter of managing things within the social media ecosystem falls under the purview of the CIO (a duty that must be explicitly stated in the governance policy). Fortunately, this function can be distilled down to a small set of explicit rules:

■ Rule 1. Nobody in the company is permitted to discuss any issue, related directly or indirectly with the firm, on any and all social media platforms. A breach of this rule should be immediately addressed and punished accordingly, up and to and including firing if necessary.

- Rule 2. The firm's executive team draws up a guidance document spelling out the mechanics of choosing a punishment as a function of the severity of the breach. This guidance document will be published as an appendix to the governance policy and disseminated throughout the firm's personnel AND exogenous supply chain partners.
- Rule 3: The firm's executive team will designate the people exempted from Rule 1.
- Rule 4: The rules of engagement between these individuals and the social media platforms shall be established by the executive team from the outset, and spell them out in clear, concise writing in a second appendix to the governance policy, disseminated as widely as the first.

That's it. With these four rules, the firm will achieve as complete a control as possible over its interactions with the social media world. One can never guarantee an ironclad protection against unsanctioned disclosures, obviously. But it is the best that can be had under the circumstances.

Note: Rule 1 will serve as sufficient deterrent if it is uniformly enforced and acted upon in a timely manner.

## Construct Management

The nature of the DCM role places it at the center of the operational readiness of the digital construct, the one that will interact most frequently with users. Data management takes center stage, aided by software management and hardware maintenance in supporting roles. Data management includes the creation of the mechanics and mechanisms underlying all datum transactions of the *plik* and *numer* sets. It is governed by concerns with access, visibility, fluidity, compatibility, operability, and recovery. Software management relates to the traditional functions of the IT department with respect to access and permissions, updates and revisions, deployment and tracking, and overhead costs. Hardware management also falls under the aegis of the IT department, according to tradition as well. However, while the IT department sits atop the software-hardware tandem, each one requires a subdivision between the *plik* and the *numer* sets to properly handle the idiosyncratic skillsets required by each one. The suggested organizational structure is shown in Figure 8.3. Note that the ensemble of processes, procedures, and standards, documenting the tasks to performed uniformly and

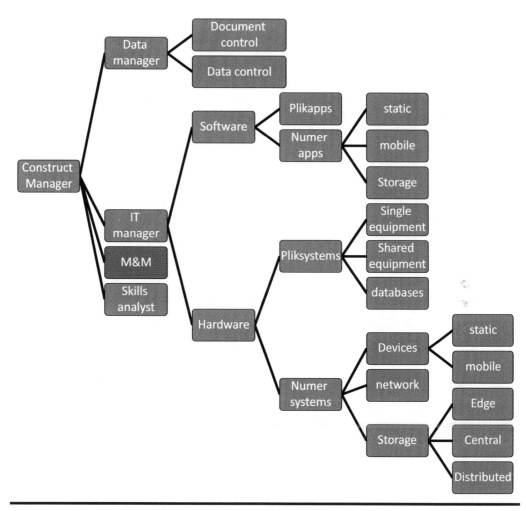

**Figure 8.3   Construct management. The organizational structure of the digital construct's management is best erected upon the physicality of the construct's features. The function "M&M" refers to the mechanics and mechanisms which document the processes, procedures, standards, and templates employed by the construct management's personnel to carry out their duties.**

specifically, falls directly on the construct manager, who may elect to add the "M&M"* function to the hierarchy, shown in gray in Figure 8.3. Other architectural approaches are of course possible for the construct management group. Nevertheless, the emphasis on the physical features of the construct, embedded in Figure 8.3, offers the best approach for aligning the various skillsets, implied by the physical features, to the functional duties of the people assigned to the management of the construct.

---

* Mechanics & Mechanisms.

## A Few Words about Document Control

The subfunction "document control", appearing under the function "Data manager" deserves a separate commentary. Document control (and its kin "document management") will likely be familiar to the reader employed by large organizations, especially those associated with the construction, engineering, consulting, legal, and government sectors. In most cases, a dedicated software application will be in place to manage the transactions involving documents between the organization and its external holobiome clients/suppliers. In most cases, the document management system will exist concomitantly with one or more file directory arrangements used by employees for internal purposes. This dual architecture is unfortunately inefficient, cumbersome, redundant, or downright confusing in most places. These inefficiencies impelled the formulation of the *plik model calls* in terms of consolidation and integration of the documents and files created by people.

> The point of the formulation is summed up simply as: abandon the duality principle then adopt a single document management paradigm for all plik actions.

The challenge for large organizations will be to resolve the existence of this duality in favor of a unique management paradigm. For the neophyte organization unfamiliar or inexperienced with formalized document management, the challenge resides in the formulation of an efficacious, *streamlined* document management policy from which the nuts and bolts of the management activities are to be performed. Once again, the reader is strongly advised not to improvise the development of the document management strategy. While the technology character of such a strategy is quite simple, its efficacy will be stymied by the myriad details that must be worked out in synchronicity. A good place to start is to consult the ISO standard 15489, parts 1 and 2. Even though the contents are, by necessity, extensive in order to satisfy all conceivable types of organizations, the broad principles suffusing the text should suffice to enlighten the reader on what is essential to design and operate a stalwart document management strategy. Beyond that, the reader will be best served by retaining the services of a competent consultancy to engineer the details of the strategy, the systems, and the implementation of the scheme.

It is important to observe that a document management system *must be designed* on a people basis. That is, it is by people, for people, of the people. Its record holdings and information assets are intended for human transactions only (if not overwhelmingly), without any direct interfaces to the other assets constituted from control, machine, and software data.

## A Few Words about Data Control

The counterpart to document control is data control, tied to the *numer* set. It deals almost exclusively with dynamic information pieces, generated continuously, periodically, or sporadically by non-human generators. The datum sets corralled by data control mechanics are not readily accessible nor modified by direct human actions. While the functionality *could be included* in the process, such functionality is meant for exception handling and forensic pursuits. Fundamentally, data control must be designed as to control, validate, verify, direct, and analyze the datum streams and information flows originating from machines. The integrity and security concerns (see *The Internet of Things* in Chapter 6) are indissociable from the design of the data control mechanics. They are, in fact, foundational to them and will overlap, in specific instances, with the mechanics implemented for holobiome oversight.

# The New Specialists

## Terra Incognita

The Symbiocene age, like its predecessors, brings its share of creative destruction to the job market. The destructive aspect was alluded to in this chapter. Now we take up the creative side and explore some of the new job types arriving on the digital scene. In some instances, the jobs are amenable to on-the-job-training and learnable as extensions of one's current skillsets (but with the mandatory caveat of advanced digital literacy). Others are true specialities in their own right, available only through external recruitment. In a few instances, a single individual can assume two or more of the jobs' functions; however, this will be possible only at the smallest of organizational scales. The likelier scenario is the one where a job type is a full-time assignment—part-time functions are best suited to third-party service contracts. Note that a transformed firm may not require all of these new jobs at once; again, the driving factor will be the scale of organization and corresponding complexity of the *numer* network. Anyone expected to interact with the digital construct (from within and even from without) will require some form of indoctrination to understand the importance of their interactions with the new digital paradigm. This indoctrination extends to the firm's exogenous linkages, especially where supply chain participants are expected to become nodal connections (inputs, outputs, or both) to the digital construct. Indoctrination in this case is not as important as the functional literacy needed by external users to plug into the digital construct (within their pre-assigned limits, obviously). Finally, the importance of a job type to the firm will vary from one to the other; for this reason, the presentation of the new jobs will be in alphabetical order.

> These new jobs will arrive on the corporate scene like foreigners on the shores of a new land. Expect some tumult, even disruptions to the old order, as they gradually find their way into the fabric of your "society".

## The Jobs

**Cloud weaver.** The *cloud weaver* designs, builds, and manages the bridges from the firm to the applications in the "Cloud". The weaver controls datum traffic to and from these applications; administers user access and licensing/maintenance issues; establishes backup and recovery plans for all data residing on the cloud (in-house, this function falls to the DCM); coordinates the joint/overlapping activities originating on the firm's website/portals; and acts as first responder in emergency cases (system crashes, data recovery, cyber attacks), in close contact with the CSO. The cloud weaver monitors the efficacy of the transactions to and from the Cloud, detects and resolves bottlenecks, and remains vigilant to the evolution of the Cloud state-of-the-art. The Cloud weaver makes recommendations to the CSO and the DCM on matters of application and service selection.

    **Data analyst.** The *data analyst* studies the datum streams supporting the firm's *model calls*. The analyst verifies the correctness of the inputs, validates the outputs, and carries out simulations and test cases to certify the performance the *model call* algorithms. The analyst also delves into the weeds of things through number crunching and analytical assessments of the statistical story told by the data. Uncovering hidden trends, deficiencies, and opportunities are critical components of this weed diving. The data analysis works alongside the *data emissary* and the *digital sculptor* to compile reports and presentation materials for management's perusal *with an emphasis on clarity of message above all other considerations*. The nature of the role is inherently mathematical and statistical; nevertheless, success with the role goes beyond number wizardry; one needs an equal understanding of the firm's strategic and tactile objectives, the business processes that go into making what the firm makes, the economic performance of those processes, and the efficacy of the processes working in symbiosis—to better detect and predict *functional* or *organizational* bottlenecks that are skewing the data. The data analyst lives under a Damocles' sword exhibited by the axiom "garbage in, garbage out". If you can't figure out whether an input is garbage or not, or whether an output is useless *even* when the input it sound, you are failing in the role in a most fundamental way.

**Data scientist**. The *data scientist* studies the underlying physiology of the data needed by the *model calls*. The data scientist formulates what its nature must be, what its genesis must be, and what coding goes into the algorithms to generate the needed data. The data scientist tells data what they must be; the data analyst verifies that they are what they were meant to be. This is the crucial task of the data scientist: the outcome shapes literally everything else that will flow from it, in terms of data intelligence, information valunomy, and numeric sentience. The data scientist is also a data analyst by nature, since pattern recognition and datum validation go hand in hand with the process of birthing data. Otherwise, you end up with a scientist disconnected with the purpose of the digital construct itself: to maximize the firm's ROI over the long run. Another component of the data scientist's mandate is to engineer ways to make disparate datum sources, data lakes, and unstructured information ponds communicate with each other efficiently. This will often mean finding ways to reduce them down to the minimal content value from which meaningful information/analyses can be extracted. A working knowledge of Hadoop, Python, and Pig, to name but three, is essential to the job. The more, the better.

The data scientist works closely with the data analyst, evidently, where validation and verification are involved. But the scientist must also work with both the *data emissary* and the *digital sculptor* to gain awareness of the importance of presenting results cogently to a *non-scientific audience.* Amazing data and stunning conclusions mean nothing if they cannot be understood in plain terms by decision-makers. Beyond those primordial activities, the data scientist sits on the front line of the mechanics and mechanisms of data capture, compilation, analysis, and presentation. The ultimate data scientist is the one who can turn data analytics into profits. The science of data, ironically, plays second fiddle in this orchestration, but still takes the form of numbers.

**Data steward**. The *data steward* is first and foremost a primordial function of the CDO, whose mandate it is to ensure that digital corporate policy is enforced throughout the organization. The data steward is concerned about the relationships between users (internal and external) and data. These relationships are defined in terms of ownership, right of access, right of creation, legal compliance, storage/archiving/vault protocols, and retention/deletion rules. The data steward and the CSO form a tandem that acts as a bulwark against nefarious intents: the data steward proposes, the CSO disposes. The data steward sits inside the information fortress while the CSO mans the perimeter wall to detect and prevent attacks. The data steward requires an in-depth knowledge of the legal and regulatory framework within which the firm operates, including, *inter alia*, laws that govern individual privacy and employees' right to privacy on the firm's premises. The steward must work hand in hand with the DCM to establish the user access protocols, monitoring procedures, and disconnect processes.

**Data emissary**. The *data emissary* plays the crucial role of ambassador—of the knowledge compiled by the data genitors (cloud weaver, data analyst, data scientist)—at the court of decision-makers. The emissary "speaks", so to speak, on behalf of all that are data, and tells their stories in the vernacular of the court. In other words, the emissary translates the genitors' complex, recondite outputs into simple, concise, and intuitive portraits (heavy on images, lean on text, nuncupative) that can be understood at a glance by a neophyte audience. This translation goes far beyond the compilation of simple charts and graphics (typically generated in Microsoft Excel); such simplicity *cannot* handle multivariate phenomena. The exact nature of the knowledge itself, conveyed by the outputs, may be esoteric and difficult to grasp holistically. Add to this mix the existence of datasets associated with different groups, independently generated but still interwoven at the corporate level, and you get a sense of the daunting challenges implicated in creating a cohesive story.

The data emissary must understand, at a working level, the nature of the outputs produced by the genitors, along with the mechanics of their production. From there, the emissary must extract the common threads running through these outputs, with the help of the genitors. Then, the emissary must figure out the best way to present these threads, which involves a definite element of artistry circumscribed by the importance of telling the story first (rather than focus on the artistry of the presentation proper).

Notwithstanding, the seemingly mundane details such as color selection, types of animation, the orientation of the presentation (portrait or landscape or other), or even the size the printed material, matters greatly to the efficacy of communication. To achieve this balance requires of the practitioner a deep understanding of the underlying mechanics employed throughout the firm to produce what it produces. Topping off everything is the imperative for proficiency in public speaking. This multifaceted role will rarely be found among the analysts and scientists, who are by nature inclined to delve deep into the weeds of things.

**Digital sculptor.** The *digital sculptor* is the technical master of rendering computer models into virtual or augmented reality. These are 3D models of buildings, plants, machines, systems, infrastructure facilities, and networks. They are typically created by architects, designers, engineers, and/or planners, who operate at the level of detail minutia bounded by physics and economics. At some point in time, these models need to be reviewed, assessed, and criticized by the people who will be called upon to make them work and make money from them. Since these people are almost never the same ones who created the models, the need arises to "tame" these models and corral them into a venue suitable for the reviewers. Often, many designated reviewers will not be able to congregate in the same room to carry out their task. The primary function of the digital sculptor is to put in place the technological and procedural elements of a presentation schema conducted virtually over the internet (or intranet), in real time, and equip everyone with the same tools and capabilities to comment as everybody else on the call.

The skillsets required by this role call for a basic understanding of the modeling tools. However, the critical skills lay with the software, hardware, and network bandwidth required to display the models simultaneously across all viewing locations *in real time* (with an absolute minimization of signal lag). The typical elements will comprise a robust video conferencing set-up allied to a clear audio system (other than the handsfree microphone of the conference room phone); a stalwart file server system able to run the simulations at all locations with near-zero lag; the VR or AR hardware with which reviewers can immerse themselves into the model; an application to capture the comments of reviewers in real time; and a web conference application to serve as virtual conference rooms. Obviously, there may be more than one digital sculptor on hand with large organizations. In such a case, one sculptor must be designated as session facilitator to run the entire show. Otherwise, the sole sculptor assumes that function by default.

It is important that this facilitator role be separate from the person running the model itself during the session. The facilitator is there to keep everyone moving in the same direction, according to the session's previously published agenda and in-session timeline. Technical exchanges will unfold among the participants as they see fit. In essence, the role of facilitator is akin to a conference chairperson.*

**Skills analyst**. This is the only new job type that does not deal with the technological underpinning of a digital transformation. The *skills analyst* either sits in the HR department or as a direct report to the DCM (as shown in Figure 8.3). The skills analyst is primarily an organizational function, in the sense of the work possibly being shared among two or more people. It can be a dedicated organizational position as well when the firm is small enough. In both cases, the job is the same: to compile a comprehensive manifest of the digital construct skillsets possessed by everyone in the firm. The skills analyst monitors the "working symbiosis" extents of each employee, conducts the performance surveys discussed earlier (see "Make the Covenants Make You Money"), and advises the DCM on the needs for training for whom, at what level, and when. The skills analyst keeps up to date with the technology advancements tagged as "important" by the CIO and devises training and development programs to get people ramped up when they are implemented (see covenants 3 and 5). The skills analyst is the primary advisor to the DCM in matters of employees' digital performance and needs.†

**The T-Ranger**. This is the only new job that extends an existing job type. The *T-Ranger* (short for Technology Ranger) is a person belonging to the IT department, whose mandate is to *keep the machines running smoothly*. This a purely hands-on, technician function, requiring of the practitioner a much higher degree of software and hardware acumen than a regular IT specialist. The T-Ranger must troubleshoot complex devices and nodal networks characterized by machine sentience. Problems originating at the device reverberate across parts of the construct's nodal networks in ways that can be non-linear, if not chaotic. The device's failure may not be readily

---

* The expression "interoperability" is a feature of a system (be it physical, algorithmic, or both) that enables it to connect to other systems (physically, controllably, and logically) such that their individual functions will operate without restriction or degradation when they are dependent on external functions for their inputs/outputs. "Plug-and-play" is the poster child of interoperability.

† A good facilitator is comfortable with public speaking and with exercising a mild authority over a group of people. If the digital sculptor is missing these traits, it will be best to assign someone else from the organization into the role (but always at the exclusion of the person driving the model).

visible or even testable and could be reacting wonkily as its edge software reacts strangely to "out-of-range" inputs. At the other extreme, high level analytical results from data aggregated from several sources within the firm could be yielding perplexing conclusions without any apparent reason to link them to systemic problems. Whatever the case, the T-Ranger will be the one called to figure out why things are coming out the way they are, outside of expectations. The T-Ranger will be especially solicited when the firm operates its own IoT. In which case, an entirely new breed of problems will come to the fore, under the heading "mobility".

## A Brave New World

These new job types are with us today, in the here-and-now, albeit still in nascent form. They exist at the bleeding edge of the digital industry; they will inevitably morph into something different over the coming decade. Indubitably, more new job types will be appearing on the horizon as the Symbiocene age progresses from disruptive tsunami to established orthodoxy. The savvy binary firm will maintain constant vigilance about what's happening on the outside, with a view to anticipating what is worth adopting or not. For the employee, the new reality of work will bring with it the expectation of constant skill development and technological malleability. Numeric literacy is no longer an option reserved for the geeky inclined; almost everyone will have no choice but to interact with the digital construct of one's employer. Changing jobs will mean adapting to a completely new digital construct; that is one inevitable drawback to the Symbiocene age. It leaves everyone with a stark choice: adapt and adopt, or stagnate and disappear.

# Bibliography

Ampere, Andé-Marie. *Essai sur la philosophie des sciences.* University of Michigan Library, 2009, 302 pages.

Babbage, Charles. *On the Economy of Machinery and Manufacturers* (original 1832). Reprinted by Cambridge University Press, Cambridge, UK, 2009, 320 pages.

Bergreen, Laurence. *Columbus – The Four Voyages.* Viking Press, New York, NY, 2011, 422 pages.

Blanning, Tim. *The Pursuit of Glory – Europe 1648-1815.* Penguin Books Ltd, London, 2007, 708 pages.

Boole, George. *An Investigation into the Laws of Thought*. Palala Press, 2015, 438 pages.

Cann, Geoffrey; Goydan, Rachael. *Bits, Bytes, and Barrels: The Digital Transformation of Oil and Gas*. MadCann Press, 2019, 209 pages.

Diamond, Jared. *Guns, Germs and Steel – The Fates of Societies*. W.W. Norton and Company, New York, NY, 1997, 480 pages.

Engels, Frederick. *The Condition of the Working Class in England* (original 1845). Reprinted by Pinnacle Press, 2017, 344 pages.

Finlay, Steven. *Artificial Intelligence and Machine Learning for Business: A No-Nonsense Guide to Data Driven Technologies*. Relativistic Books, London, UK, 2017, 150 pages.

Freeman, Joshua B. *Behemoth – A History of the Factory and the Making of the Modern World*. W.W. Norton & Company, New York, NY, 2018, 427 pages.

Harari, Yuval Noahi. *Sapiens: A Brief History of Humankind*. Harper Perennial, 2018, 464 pages.

Harris, Karen; Kinson, Austin; Schwedel, Andrew. *Labor 2030: The Collision of Demographics, Automation and Inequality*. http://bain.com/publications/articles/labor-2030-the-collision-of-demographics-automation-and-inequality.aspx.

Higginson, Matt; Nadeau, Marie-Claude; Rajgopal, Kausik. *Blockchain's Occam Problem*. McKinsey & Company, January 2019. https://www.mckinsey.com/industries/financial-services/our-insights/blockchains-occam-problem.

Holmes, Arthur. *The Age of the Earth*. Forgotten Books, 2017, 222 pages.

ISO standard 15489-1. Information and *Documentation - Records Management, Part 1 – General*.

ISO standard 15489-2. Information and Documentation - *Records Management, Part 2 – Guidelines*.

Keays, Steven J. *Investment-Centric Project Management: Advanced Strategies for Developing and Executing Successful Capital Projects*. J. Ross Publishing, Plantation, FL, 2017, 419 pages.

Keays, Steven J. *Investment-Centric Innovation Project Management: Winning the New Product Development Game*. J. Ross Publishing, Plantation, FL, 2018, 309 pages.

Kelsey, Todd. *Surfing the Tsunami – An Introduction to Artificial Intelligence and Options for Responding*. Todd Kelsey Publisher, 2018, 188 pages.

Kranz, Maciej. *Building the Internet of Things – Implement New Business Models, Disrupt Competitors, Transform Your Industry*. John Wiley & Sons, Hoboken, NJ, 2016, 272 pages.

Mar, Bernard. *Data Strategy: How to Profit from a World of Big Data, Analytics and the Internet of Things*. Kogan Page Limited, London, 2017, 200 pages.

McLuhan, Marshall. *Understanding Media: The Extensions of Man*. MIT Press, 1994.

Morison, Samuel Eliot. *Admiral of the Ocean Sea: The Life of Christopher Columbus*. Atlantic/Little, Brown, Boston, MA, 1942. Reissued by the Morison Press, 2007.

Rifkin, Jeremy. *The Third Industrial Revolution: How Lateral Power is Transforming Energy, the Economy, and the World*. St-Palgrave MacMillan Publishers, New York, NY, 2011, 304 pages.

Rogers, David L. *The Digital Transformation Playbook.* Columbia Business School Publishing, New York, NY, 2016, 278 pages.

Rossman, John. *The Amazon Way on IoT: 10 Principles for Every Leader from the World's Leading Internet of Things Strategies.* Clyde Hill Publishing, Washington, DC, 2016, 168 pages.

Russell, Stuart. *Artificial Intelligence – A Modern Approach.* Pearson Education Limited, London, 2015, 1164 pages.

Sacolick, Isaac. *Driving Digital – The Leader's Guide to Business Transformation Through Technology.* American Management Association, 2017, 283 pages.

Schwab, Klaus. *The Fourth Industrial Revolution.* Crown Business, New York, NY, 2017, 192 pages.

Sills, Franklyn. *Foundations in Craniosacral Biodynamics, Volume 1: The Breath of Life and Fundamental Skillsets.* North Atlantic Books, Berkeley, CA, 2016, 424 pages

Sloman, Steven; Fernbach, Philip. *The Knowledge Illusion: Why We Never Thing Alone.* Riverhead Books, New York, NY, 2017, 304 pages.

Smith, Adam. *An Inquiry into the Nature and Causes of the Wealth of Nations.* The University of Chicago Press, 1976, 1152 pages.

Smith, Adam. *The Theory of Moral Sentiments.* Digireads.com Publishing, 2010, 238 pages.

Tapscott, Don; Tapscott, Alex. *Blockchain Revolution: How the Technology Behind Bitcoin Is Changing Money, Business, and the World.* Portfolio/Penguin, 2016, 432 pages.

Vigna, Paul; Casey, Michael J. *The Age of Cryptocurrency: How Bitcoin and the Blockchain Are Challenging the Global Economic Order.* Picador, New York, NY, 384 pages.

Westerman, George; Bonnet, Didier; McAfee, Andrew. *Leading Digital – Turning Technology into Business Transformation.* Harvard Business School Publishing, 2014, 256 pages.

Wiener, Norbert. *Cybernetics - 2nd Edition: Or Control and Communication in the Animal and the Machine.* The MIT Press, Boston, MA, 1965, 212 pages.

Windpassinger, Nicholas. *Internet of Things: Digitize or Die: Transform Your Organization: Embrace the Digital Evolution: Rise above the Competition.* IoT Hub Publishers, New York, NY, 2017, 284 pages.

Zola, Emile. *Germinal.* Penguin Books, London, 2004, 592 pages.

# Chapter 9

# Project Digital

*The mounds of data, information, and knowledge created during the delivery of a major capital project amount to exactly zero value in terms of what the owner bought.*

This chapter is a case study about performing a digital management transformation at a micro-scale, encapsulated by mechanics of executing an industrial capital project within a self-contained binary framework separate from the remainder of the firm's business structure.

## Capital Projects as Test Cases

### A Different Kind of Pilot Transformation

The world of capital projects offers a different path to conduct a pilot digital transformation, even while the organization remains analog in all other aspects of its operations. Capital projects build large things like airports, bridges, buildings, mines, pipelines, power plants, ports, roads, solar plants, wind farms, etc. They mobilize enormous investments in money, time, materiel, and people, expended over months or years, and yield physical assets with large footprints attracting perennial operating costs covered by profits, revenues, or taxes. By this definition, one could extend the franchise to other investment-intensive design endeavors to develop new technologies, machinery, and even software. Software projects proceed in the virtual, ethereal world of codes and algorithms to develop new codes and new algorithms. Code begets code; software begets software. For all other kinds, code and software beget physical assets, through which the human variable comes into intimate play via supply chains, manufacturing, and construction. Hence the seminal difference between software projects and all others: for the latter, the development process will be dominated by the complex interactions among the supply chain participants (what, in *investment-centric project management* parlance, is conveyed by the *project ecosystem*, or PECO for short).

We can probe the complexity of projects through the lenses of information theory. It becomes clear that a capital is nothing but a microcosm of a digital management transformation. The naia, created during the development process,* is to a project what scaffolding is to cathedral building: the essential lattice without which an asset cannot be erected. It emerges over time as a nodal network (cf. Chapter 2) spread over three information layers: the *framework, development*, and *monetization* layers. The first layer is made up of the plans, schedules, budgets, strategies, and policies formulated by the asset owner's organization to govern development. The second layer occurs throughout the development works and grows organically and exponentially as the state of the asset becomes increasingly detailed. Evidently, this state will vary over time from requirements to specifications to design to fabrication, and finally, monetization (a process called *the esemplastic key*, described in Chapter 12 of *Investment-Centric Project Management*).† The third layer begins at the same time as the asset comes online. It will forever keep growing as it accumulates data from operations.

---

* The term "development" abides by the definition explored in Chapter 2 of *Investment-Centric Project Management*. A project's development spans two sequential phases: *conceptualization* and *realization*. The former begins with the initial concept of the asset and ends when the construction basis is completed (from which the asset can be constructed). The latter begins with construction planning and concludes with the operational validation of the asset's profitability performance.

† The esemplastic key was described in Chapter 12 of *Investment-Centric Project Management* as the fundamental means of sequencing the work that goes into turning an idea into reality. The key spans the design lifecycle of that idea from concept to operation, and establishes the information flow between the six developmental steps involved: FR > FS > DS > PS > B > M, where FR = functional requirements, FS = functional specifications; DS = design specifications; PS = procurement specifications; B = build; and M = monetize (operate to generate revenues). Each step in the sequence is analyzed through the unit transformation process. See also Figures 12.2, 12.3, and 12.4 in Chapter 12.

## *The Opportunity*

It should be clear by now why capital projects are such fertile grounds for a digital management transformation: the entire lifecycle of these assets can be regarded as an epic trek through a valley of data seemingly never ending. The end goal is a *binary project delivery framework*. The opportunity being *justified* stems from the morass in which capital projects find themselves from current project delivery practices. The scaffolding was the naia by the end of the project is effectively discarded by the *monetization layer*, other than static, passive remnants of drawings and datasheets stored back at Corporate, or in some paper archive collected offsite. In the digital paradigm, the scaffolding would automatically morph into the digital *foundation* of the monetization layer. In reality, nothing of the sort happens. The asset begins life from a nearly blank slate when it comes to its *monetization* layer.

Sadly, even if one were to plan to morph the project's first two layers into the asset's *monetization* layer, the initiative would be stillborn. For the defects and inefficiencies of the *plik* and *numer* sets (cf. Chapters 5 and 6) employed during development render the project's naia digitally impotent (see Table 9.1 for example). The profit-killing shortcomings of both sets propagate and multiply like locusts during development. Having a document saved electronically does not make it *digitalized*, in the sense of nodal network integration. It merely makes the document *accessible* passively and statically, provided of course that it can be located in the first place (which is never a given in a non-transformed organization).

Aggravating the cost torments is an absence of any *numer* and intelligence solutions (see Chapter 6) in the day-to-day execution of the development work. And to add insult to injury, the design of the asset will avoid imbuing the design of the asset with ROI-enhancing compatibilities with potent analytical *numer* solutions described in Chapters 2 and 6. Frustratingly, current asset design and operating philosophies actively preclude such considerations, thereby condemning the future assets to an analog purgatory with nary a hope of ever joining the Symbiocene age *economically*.

> Capital assets will continue to be orphaned from the digital world until the project delivery strategy itself makes the commitment to inscribe into the DNA of its developmental methodology the obligation to digitalize the asset as well as the journey that will create it.

**Table 9.1 The Plik Warts "All Is File" Sums up the Character of the Information Ecosystem Created by a Typical Project Delivery Schema**

| *Feature* | *Wart profile* |
|---|---|
| The document is the transaction | Whether as a text or spreadsheet, the document becomes the transaction, and proliferates without bound or control as the product of participants and the number of revisions issued. Standardization and content management do not exist beyond each instance of the document. The document is a digital orphan, untethered to any information set or database. |
| The contents are indissociable from the document | Because the document is a singular instance of a transaction, its contents are equally unique to each document, even when multiple instances of a document could be replicable from an automated template. Consequently, content standardization across a multiplicity of document types with common contents is impossible. Contents become non-searchable across an information set and cannot be optimized or updated uniformly across all instances of documents containing them. The contents are, like documents, digital orphans. |
| Tracking and metrics are stranded | Lists, tables, and other like contents, intended to be regularly tracked and updated as new information comes to light, are condemned to manual transactions (via Excel for example), without any possibility of automation across the project's documentation ecosystem. Higher order analyses and AI assessments are out of reach. The update data are static, forever orphaned, and outside the reach of any optimization effort. |
| Design derivatives are orphaned | Information instances derived from design systems, such as drawings from 3D models, and material lists from drawings, exist only within the closed information ecosystem anchored by the model. The derivatives are static, not given to automatic updates via dynamic linkages to the models when modified, and completely invisible and unreachable by any other transaction mechanism existing within the project's information ecosystem. Such derivative information instances are forever stranded on their own data island. |
| All is file | Every piece of datum, information, knowledge, and intelligence engendered during project development is captured and stored as a singular information instance, usually bereft of any dynamic linkages to other instances of relevance. Every piece is stored in various databases and file servers, but nothing is ever integrated into a nodal network that preserves the dynamic linkages and permits higher level analyses and algorithmic assessments. The project's information landscape yields a scattering of hundreds or thousands of isolated data islands. |

*(Continued)*

**Table 9.1 (Continued)  The Plik Warts "All Is File" Sums up the Character of the Information Ecosystem Created by a Typical Project Delivery Schema**

| Feature | Wart profile |
|---|---|
| Models never make it into the asset | The multitude of 3D models and their design derivatives never become the digital foundation of the future asset. They are effectively set aside once the asset is built, never to be integrated into the asset information set. Effectively, the asset's information ecosystem shuns everything that was ever created during development. |
| Primordial and developmental information sets are abandoned | In parallel with the fate of the models, all of the data, information, knowledge, and intelligence instances accumulated during the development works end up either on shelfs or servers back at the owner's project office, never to be integrated into the information ecosystem of the asset. Effectively, all the money and effort invested into the creation of these information instances are tallied as ersatz sunk costs invisible to the operating asset. |
| It is a spreadsheet world after all | Despite the existence of potent software applications employed throughout project development, the preferred tool of management and progress oversight remains the stoic, static, and error-prone spreadsheet. The entire project development process is thus inaccessible to stalwart analytical processes, process automation, and information flow efficiencies. Spreadsheets condemn the project management organization to a precarious existence perpetually spent in reaction to sudden events, rather than anticipate what lies ahead. The act of management unfolds at the speed of belated inertia. |
|  |  |

*Note:* When all is file but without integration, the result is a massive archive of worthwhile information instances that cannot be brought into digital practice. Once the project is done, their usefulness is also done.

That is the quintessence of the opportunity offered to the organization intent on being progenitor or sponsor of even one single capital project. The beautiful thing about it is the fact that this opportunity exists today, regardless of the digitalization state of a project delivery organization. Obviously, the impact of a transformation to a capital project will be proportional to its stage of development;* the earlier, the greater the positive impact to the bottom line. If the development is nearing the end of conceptualization

---

* The overall development sequence of a capital project comprises eight stages, together called *the ICPM lifecycle stages* in *Investment-Centric Project Management*, Chapter 12.

(see Figures 9.1 and 9.2), it is best to continue as planned, rather than undertake the transformation.*

Despite the existence of the opportunity, one may still ask why bother with the digital ordeal in the first place? After all, the world of capital projects has managed to get by forever without a digital playbook. That may be true, but the sobering track record of failures across industries the world over would seem to indicate that not all is well in the traditional project delivery world. When 50% of projects valued above $100M, and 70% above $1B bust their budgets, blow their schedules, and subceed[†] their performance expectations, one would expect the discipline to seek all the help it can get. Everything else being equal, undertaking a capital project digital management transformation is worth it for the following reasons:

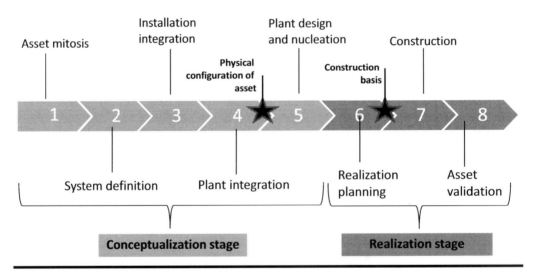

**Figure 9.1 The asset development sequence. Development unfolds in two phases, conceptualization and realization, jointly divided into eight lifecycle stages matching the fundamental definition of a project as "the development of a profitably performing asset (PPA)".**

---

* The expression "project delivery" encompasses all the organizational aspects of a capital project, one of which is, naturally, execution—to carry out the activities of development in an ordered, managed manner such that the profitably performing asset is obtained. Execution is complemented by other, equally important elements such as concept formulation, economics calibration, commercial viability, initiation and funding, supply chain management, framework oversight, and full-up operations. Among this field, execution begins when commercial viability is confirmed and ends before full-up operations.

† The neologism "subceed" was first coined in *Investment-Centric Project Management* as the antonym of the word "exceed". To subceed is to go below and beneath expectations.

| Asset Mitosis (1) | System definition (2) | Installation integration (3) | Plant integration (4) |
|---|---|---|---|
| Define the overall asset requirements. High-level targets are assigned. Identify the PECO objectives, investment message. Distill plant into primary installations. | Define the functional specifications for the primary systems. Define the IBL requirements of the secondary systems. Identify OBL commands. | Integrate the primary and secondary systems of each installation into functional networks. | Integrate the primary and secondary installations into the pant's functional network. Develop the construction requirements. |
| **Plant design (5)** | **Realization planning (6)** | **Construction (7)** | **Asset validation (8)** |
| Complete the design of the entire plant. Derive the construction specifications. Develop all Phase 4 FS into DS. Finalize all command and control details of the plant. Commence the nucleation work. | Assemble the *construction basis*. Develop the plans to construct the asset. Execute the permit plan. Award all required contracts. | Initiate the site preparation works. Begin construction. Continue until the plant is ready for commissioning. | Enter operational readiness. Execute start-up and commissioning. Complete personnel training. Operations take-over. Verify plant's performance. Modify as required. Collect PAMs. Warranty claims and hold-backs are resolved. Complete close-out. |

**Figure 9.2 The lifecycle stages. Conceptualization comprises that go on before a shovel even breaks ground (aka the "pre-shovel works"). Realization spans everything afterwards, until such time as the profitability performance of the asset has been properly assessed.**

- Reduce delivery costs and timelines across the board
- Eliminate unforced errors, omissions, reworks, workarounds, and miscommunications
- Dramatically cut project overheads while simultaneously increasing throughout and productivity
- Increase quality, repeatability of process, accuracy of outputs, and certainty of deliveries
- Eliminate the warts, bottlenecks, and profit-killing shortcomings of analog *plik* and *numer* frameworks
- Eliminate paper, impotent spreadsheets, and the proliferation of uncontrolled versions
- Enable the entire supply chain to work from a single, unifying naia foundation
- Implement true visibility, transparency, and accountability among all project participants
- Achieve true and genuine instantaneous progress tracking and health assessment
- Banish useless, redundant, or efficiency-killing processes and procedures

- Strengthen and solidify the perennial ROI performance of the realized asset
- Make more money, save more costs, and get things done faster

## *First, the Empire Must Fall*

The ultimate effect of a non-digital project delivery philosophy can be summed up as follows:

> The project's naia will evolve into a morass of disconnected, disjointed information instances incapable of being integrated into an asset's planned digitalized philosophy.

The consequence of this effect is to end up with a naia that ceases to matter to the asset. The scaffolding is effectively dismantled for good and vanishes from view. All the money spent on development becomes sunk costs, without future value—an abhorrent accounting outcome if ever there was one. None of these consequences needs to be suffered, however, if the obdurate *status quo* is denied the stranglehold that it currently holds over project delivery practices. The analog empire must fall.

The attack plan follows the logic expounded in previous chapters for making the case for a digital management transformation. The salient strategic objectives will include:

- A declaration by the owner's leadership that the asset shall be digitally designed and operated
- A mandate granted by said leadership to digitalize the project delivery organization
- The establishment of a digital management transformation team to carry out the digitalization
- The formulation of the *model calls* for the asset
- The formulation of the *model calls* for the project delivery organization
- The execution of the digital management transformation

The salient tactical objectives that follows will include:

- Define and roll-out a digital policy (cf. Chapter 8)
- Engineer a bespoke digital construct specific to project delivery
- Abandon paper and spreadsheets

- Adopt advanced *plik* and *numer* solutions
- Port the delivery framework's nodal network onto the Cloud
- Measure, track, and quantify progress through visualization channels (instead of orphaned reports, spreadsheets, and slide presentations)
- Anchor the nodal network to the physical configuration of the asset

## About This Business of Asset Physical Configuration

The typical project delivery organization will maintain three file frameworks in parallel: the project folder, to file all matters of data, documents, and information (memos, budgets, schedules, contracts, purchase orders, datasheets, calculations, manuals, travel requests, etc.); the document management system, to store all deliverables produced by all supply chain partners (drawings, technical data, inspection reports, permits and licenses, etc.); and the application systems, comprising design software, engineering simulations, planning and scheduling, ERP and MRP, timesheets, and other specialized software. Each framework will have its own, usually independent, file structure, without any connectivity or linkages to the other ones. The project folder and the document management system could overlap in certain areas but will still rely on a storage architecture following a *labor taxonomy*,* such as the example shown in Figure 9.3. Now, imagine the following scenario involving the construction of a pipeline, with the project folder mirroring the architecture of Figure 9.3. The pipeline, normally buried, regularly comes up to the surface (at a location called a surface lease) to connect to various pieces of equipment for monitoring and/or processing requirements. The construction of the pipeline is by Firm A, while that of the surface lease is by Firm B. During construction, the owner receives a request from a potential client to add a connection for that client to bring his own pipeline flow into the mainline being constructed. Drawings and specifications are prepared for the purpose. The question is, where in this folder structure should the drawings and specifications be saved? In project management, in design, or in construction management? In the engineering subfolder, the Firm A or Firm B contract folder, or under the new client folder? Unfortunately, it does not matter in the end where these documents are filed because, outside of the person who filed them, there is literally no way for anyone else to logically and intuitively navigate the folder's contents to get to those files.

---

* In Chapter 4 of *Investment-Centric Project Management*, the taxonomy is referred to as the "labor space", while the physical features of the asset are called the "feature space".

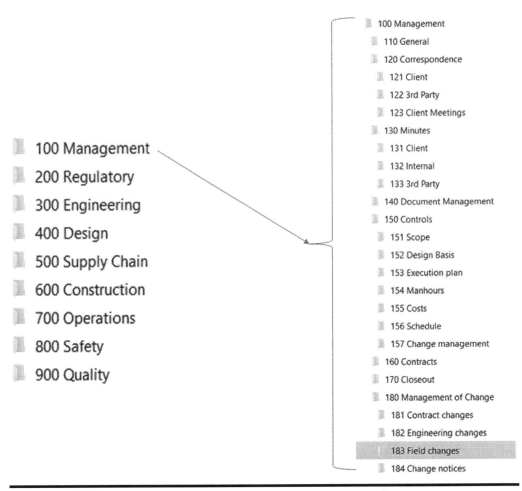

**Figure 9.3 The traditional project folder structure. The most common schema for organizing project files relies on labor as the organizing principle. The schema is near useless in organizing the contents of the asset that is being developed. Note for instance the redundancy between folders 157 and 180.**

This example goes to the heart of the argument in favor of the physical configuration of the asset (explored at length and justified in *Investment-Centric Project Management*, notably in Chapters 4, 12, and 13). The configuration, named *feature space*, is the only one capable of capturing holistically the interactions, interfaces, and linkages that will arise in the course of the project's development. In summary, the highest level of the architecture is the asset. Next level includes the *installations*, which are divided into *systems*, and each system into *components*. Hence, the basic configuration comprises four levels. The levels themselves are thrice divided: *primary* (that which exists in support of revenue generation), *secondary* (what enables the primary to fulfill its functions), and *tertiary* (what makes the primary-secondary tandem to run). Then, and only then, does the architecture introduce further divisions driven by labor or activity or task (usually performed by humans or organizations). In our pipeline example, the asset is the entire pipeline from end to end. The surface leases are installations. The installation will include such systems as pumps (for liquids) or compressors (for gases), RTU/control buildings, and storage tanks. The tank system, to mention but one, will include such components as the shell, the level gauge, the refilling pump, and a thief hatch to prevent overpressure. When the surface lease is meant for fluid moving (pump or compressor) or flow heating, the installation is at the primary level. If the lease is meant as an emergency shutdown station (to isolate the pipeline when a leak is detected), the level is secondary. The access road and the fence around the lease are at the tertiary level.

# The Building Blocks

## *The 3D Kernel*

The importance of the asset's physical configuration lies in its purpose as organizing principle. It becomes the cornerstone of the project's nodal network, to which every future naia instance will be attached; otherwise, the unbounded nature of these instances inevitably leads to an accumulate mound of disparate, unreachable, or even invisible records lost somewhere on some server or hard drive. Nevertheless, this organizing principle is unworkable on its own without it being given substance by means of aggregating and disseminating the naia instances both efficaciously and intuitively. We need, in other words, an interface between the principles and the naia instances governed by them. This interface is called the *3D Kernel.*

The 3D kernel is the portal through which users access and query the project's naia. The kernel sits above the multitude of applications that generate and store the naia instances (schematically illustrated in Figure 9.4). The kernel aggregates the instances into a virtual model which can be plumbed and queried by a user for whatever information. The virtual model is inherently intuitive to the user, thus granting the transaction act between user and naia a potency and efficacy that is unmatched by any other means. Significantly, the user has no need to understand the hidden filing structure and organization of the naia; this understanding is

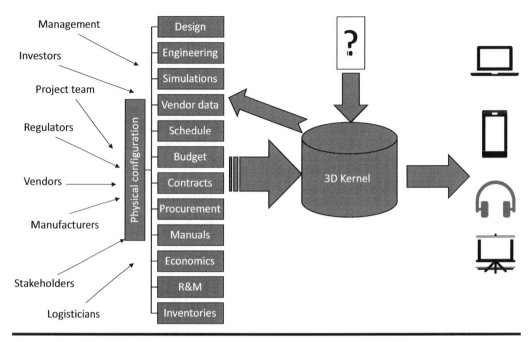

**Figure 9.4  The 3D kernel ecosystem.**

embedded into the GUI functionality of the kernel. In this respect, the kernel operates much like Hadoop (which could indeed be its algorithmic engine). However, the kernel is more than Hadoop in an operational setting, owing to its capability to provide real-time, on-the-fly progress and execution metrics in dynamic visualization schemes (engineered *a priori* by the *data emissary* and the *data sculptor*—see Chapter 8). While the "access-by-model-navigation" is powerful on its own for purposes of project information transactions, it is the "visually dynamic progress and metrics" capability of the kernel that yields the greatest bang for the project delivery buck. With it, whoever on the project bears a direct accountability role can genuinely manage his/her execution mandate proactively,* with the ability to spot potential problems and anticipate their impact before they take place.

The 3D kernel is NOT, on the other hand, a generator of naia instances. Instance generation occurs outside of it, by whatever software is in use by project participants. The separation of powers between "aggregation" and "generation" is an inevitable consequence of the reality of capital

---

* Proactive management is a seminal success factor to anyone managing some aspect of the project's development. See Chapters 3, 4, and 15 of *Investment-Centric Project Management,* and Chapters 3 and 13 of *Investment-Centric Innovation Project Management.* Managing after the fact, in reaction to a foreseeable event that already happened is not project management: it is journalism.

project delivery. It is never possible for an asset owner to insist on imposing a specific suite of software applications upon the project team and the myriad supply chain partners that will eventually partake in the development works (the expense would be beyond prohibitive in any case). The kernel needs only to be able to establish a communication channel with whatever application it intends to extract information and to translate the incoming datum stream into its own "language". This minimization principle extends to the visual presentation of the virtual model as well: it needs only be *visually* faithful to the underlying, much more complex model created by the specialized application. If a user wishes to modify the model, and run what-if scenarios for example, the 3D kernel is not the tool to use; one must go back to the model's generator. The 3D kernel is thus unidirectional, from application to display. It cannot reverse the direction of this datum transaction. Notwithstanding the algorithmic hurdles involved, the reason for this unidirectionality is obvious: to enforce the integrity of the naia, which otherwise would be hopelessly compromised almost from the outset.

## MAKING THE PROJECT MAKE YOU MONEY

Once the *monetization* stage of the development is completed, the project will be completed. The size of the naia will have reached its maximum extent. Coupled with the 3D kernel, the two will effectively form the complete DNA and digital build-up of the asset. In traditional project management, this point also marks the moment when all these information assets are abandoned or simply ignored. Within a *binary project delivery framework*, the naia-3D Kernel duo morphs from project lattice to asset's digital foundation. From that point forward, the operationally generated naia of the asset will be appended to the project naia. Clearly, the expanded naia will forever grow in volume as long as the asset incurs costs. This richness of data and wealth of extracted knowledge will add to the intrinsic value of the asset, both operationally and commercially. The binary transformation of the project delivery framework will have come full circle; the big dataset constituted by the asset naia will carry its own investment value, with every potential for further monetization independently of the asset' revenues. The naia, in other words, will have expanded the reach of the asset's long-term ROI performance.

## Visualization Templates

The importance of transformative *plik* solutions bears repeating at this point: the old ways of document creation, perpetual wheel re-invention, and spreadsheet proliferation are incompatible with an efficient binary project delivery framework. Dynamic reporting, especially of changing progress metrics, cannot proceed from spreadsheet files, static drawings, orphaned purchase orders, or shelved plans and procedures. The prohibition extends to the presentation of aggregated contents by the kernel: the best delivery medium is the visual, laconic in text and rich in imagery and animation, and always superimposed on a background comprising the virtual model. The presentation itself must allow the viewer to query the kernel through what is displayed, without cumbersome menu drilling or repetitive query submissions. The *data emissary* plays a pivotal role in the creation of the various visualization templates that will be embedded in the GUI of the 3D kernel, from the outset and as time goes by. Reviews of the model itself will entail additional technical mechanics falling under the purview of the *data sculptor*. Ideally, these templates will be created in the earliest stages of the project, when the naia volume is small. Others will come up over time, when users voice their requirements for different means of presenting the data. In all cases, templates must be created by the *data emissary* and/or the *data sculptor*. Never let project people improvise themselves into these roles.

## Autonomous Progress Metrics

During a project's development, the project management team will always be concerned with the rate of progress of the conceptualization and realization works. Traditionally, tracking a project's progress has relied on regular updates to the master schedule, coupled with monthly reconciliations of the accrued costs against the budget. The entire exercise is reactive, passive, and driven by *manual* updates captured from *subjective assessments* voiced by the pertinent accountable parties. The subjective approach is an unavoidable artifact of the *labor taxonomy* underlying the structure of the project (discussed earlier in this chapter). Migrating the project structure to the *feature space* from this physical configuration of the asset allows one to track the progress of the work via subroutines built in the 3D kernel. Past project experiences will yield a treasure trove of historical data from which the current project's task durations and costs can be budgeted with data analysis applications. Work sequencing can be automated as well by AI-type algorithms capable of learning from past projects. In summary, the planning, scheduling, costing, and sequencing of the development work can be automated from the outset, and tracking performed in similar fashion based on the actual state of progress of every deliverable under development. The subjective human assessment can be completely taken out of the process—which is essential to any hope of performing real-time progress tracking.

## Construction Tracking

The autonomous tracking principle applies to all activities associated with fabrication and construction, which are traditionally difficult to monitor precisely. The tracking process first requires the 3D model to be configured into work sequence layers which track the state of completion of each bit and byte by their physical condition. The sequencing can also be subjected to machine-learning algorithms, based on past experience and known scheduling constraints (such as late component deliveries). This planning is then coupled to robotized progress surveillance (by drone or laser scans, for example), which are able to map, in real time, the digitalized picture of the physical work against the control work sequence. Once again, the subjective human element is eliminated, yielding an unbiased assessment of the state of progress of the work. The approach is applicable across the scale of a project's configuration details, from the simple manufactured component to the entire construction worksite. The only limiting factor is the robotized machinery deployed to digitally capture, in real time, the daily condition of the work.

## Intelligent Tracking

The automated tracking metrics process described above involved one algorithm to quantify the expected duration, costs, and resources of a sequence of works, and one robotic system to create an instantaneous digital picture of what's been done and what's missing. We can add a third algorithm, also belonging to the AI club, to perform real-time assessments of the likelihood of completing the work in accordance with the targets. The assessment can uncover hidden trends that point to late completion or cost overruns, in case the sequence of the work is going off the rails or is being hampered by delays (labor shortages, missing parts, late material deliveries, test failures, etc.). These assessments are especially insightful on construction sites, where large work numbers, vehicle traffic, and large equipment movements could be creating unforeseen bottlenecks or catastrophic alterations to the on-site sequence of work or deliveries (for example). Crowd control, weather and natural effects, or regulatory delays exemplify the kinds of exogenous constraints belonging to the PECO that are unpredictable yet consequential. In such cases, the assessment algorithm can be employed to figure out the changes to the original work sequence that will minimize the hits to the schedule, the budget, and/or the performance of the asset.

## Performance Assessment Metrics

Traditional project management relies on so-called key performance indices (KPIs) to track the progress of the development works and quantify the earned value accrued to the owner over time. These metrics will indicate, from a 30 000 feet perspective, how the budget is being spent, but they are notoriously ineffectual in assessing *why* the targets set for the execution metrics have not been met at a given moment. Furthermore, the metrics yield zero insights on the reasons why an execution strategy is faltering, why a sequence of work is not working, or why errors keep cropping up here and there with nefarious effects. In fact, KPIs work *against* the principle of continuous improvement, by pointing out the obvious (you are late, you are costing more than you should) and focusing exclusively on stopping these undesirable excursions beyond the expected norm. The fact that KPIs can be readily compiled in real time within a *binary project delivery framework* does not support the argument for doing so in the first place. Indeed, as argued in Chapter 10 of *Investment-Centric Project Management*, the KPI approach creates the illusion of management oversight at the expense of the truth of a project's reality in the moment. A better metrics basis exists, one that is naturally suited to the automated progress tracking discussed above. It is called the *performance assessment metrics* (PAM) methodology. The reader is invited to consult the aforementioned Chapter 10 to explore the method and the means of calculations.

## Reality Visualization

Augmented reality (AR) and virtual reality (VR) processes have an important *economic* role to play during the conceptualization stage. Their value proposition at the technical level is self-evident: they allow project stakeholders other than the engineers to participate directly into the evolution of the asset's design. Model reviews, constructability reviews, hazard assessments,

operability assessment, construction sequencing, and even personnel training will be vastly improved by participants who can fly through the virtual
rendering of the asset (via the 3D kernel or dedicated virtual meeting room
settings). Yet, the greater valunomy to the project is to its bottom line. AR
and VR tools completely eliminate the prohibitive costs associated with
travel, accommodation, and lost production for out-of-office participants. The
complications that come with trying to schedule a review to match everyone's availability are correspondingly mitigated; it is far simpler to line up
everyone for a half day or full day review session without incurring lost time
from travel. Furthermore, the elimination of travel throws open the door to
all pertinent participants, who can just plug into the presentation system
with the same access as everybody else, in real time, without limits.

> Reality visualization is thus thrice beneficial to the project: 1)
> by solidifying the solidity of the design before it is constructed
> (increased asset ROI); 2) by catching errors, omissions, or configu-
> ration conflicts before they are baked into the design; 3) by maxi-
> mizing the project's access to the entire corporate expertise; and 4)
> by doing all of this while cutting budget costs and shrinking work
> schedules.

## Materiel Management

In the capital project context, materiel management is concerned with the
visibility and traceability of every *physical* item that is bought during development for the purpose of fabricating, constructing, installing, and operating
the future asset.* Some materiel will be bought once and never again revis-
ited (concrete, road asphalt, water removal during construction), while others will require continuous monitoring while in operation (essentially, any
equipment needing maintenance or replacement due to wear and tear). The
profitability of an asset is directly affected (good or bad) by the incurred

---

* Software, control programs, and algorithms of one sort or another will also be integral to the
  scope of procurement of a project. However, these numeric products are usually better served
  under the aegis of the digital construct manager (see Figure 8.3). Software management is quite
  distinct from physical materiel management and involves different mechanics and mechanisms of
  lifecycle management than their physical brethren. For instance, software is not typically tracked
  in terms of reliability, maintainability, spares, and operating costs (beyond yearly maintenance fees
  and user license renewals).

costs derived from such materiel management mechanics. Obviously, the efficaciousness of those mechanics is entirely affected by the *traceability* of the materiel and materials so managed. One can ill manage a component, for example, that doesn't show up on anyone's maintenance plans. The term *traceability* is key to everything and comprises:

- Visibility: whether or not a managed piece exists or not as an accessible record within the naia. If not, the piece is effectively invisible to the organization but present in the asset's operating scheme; hence, materiel management costs can never be budgeted and will come as a negative surprise to the bottom line.
- History: the details of a piece's antecedents, from concept to specification to procurement to delivery to installation. A large number of materials and equipment will come with bespoke documentation (part number, serial number, PO number, vendor information, fabrication quality control, parts lists, test results, calibration, composition and analyses, etc.), which must be kept throughout the piece's lifecycle and often maintained or updated (think service bulletins, recalls, warranty actions, etc.). Without history, the piece becomes an orphan.
- Reliability: the on-going measurements of a piece's failure behaviour over time. This failure behaviour includes uptime ratings, mean-time-between failures (MTBF), mean-time-to-failure (MTTF), calibration drift, performance drift, etc. Reliability tells the asset manager how often a piece will need to be looked after, or replaced, and the operating budget needed to keep it running.
- Maintainability: the level of effort (scope and frequency) to keep a piece operating, including the costs associated with those efforts over time. Maintainability also covers configuration management, which keeps track of the changes in the make-up of equipment, machines, or processes. Configuration management tells the asset manager about the kinds of replacement parts that will be needed at what moments in time, in what quantities. It also tells the asset manager about the compatibility of a part in relation to another: which part can replace which older ones, which old parts are no longer used or sold, which new parts will be coming up for release. And of course, all naia instances implied by maintainability are included.
- Obligations: some equipment or processes are subject to regulatory oversights over operating limits (noise, exhaust emissions), reporting requirement under failure (leakage of noxious gases in a plant), and

throughput metrics (net flow volumes by a pipeline). These "obligations" become a mandatory principle of operation by the plant and must be at all times known, tracked, and reported compliantly.

The mechanics of materiel traceability* have been around for a long time and embedded in the niche market called *building information management* (BIM). It is regular feature of the commercial real estate world. It has yet to make deep inroads into non-building capital projects. Hence the opportunity to include traceability into a project's naia. Once linked to the 3D kernel, it will become the de facto BIM foundation of the asset, when the latter comes into operation, under the aegis of its *digital foundation*.

## Storage Strategies

A binary project delivery framework does not preclude the traditional reliance on servers, databases, and PC hard drives for storing naia instances. On the contrary, these common systems must remain in use; otherwise, the upheaval costs to the organization will kill the transformation from the get-go. A document management system, complete with processes and procedures, is also advised to remain operational. Nevertheless, with or without the 3D kernel, there is a requirement for the project to make the naia holdings available to all project participants via the internet (or the Cloud). The underlying system architecture for this access can, at a minimum, be based on services like Sharepoint, Box, or Dropbox (to name but a few); they are not, however, suitable to a 3D kernel paradigm. The better approach is a virtual project space, located in the Cloud, and owned by the project management team.

The importance of the virtual project space becomes paramount once the asset achieves operational readiness. At that moment, the naia-3D kernel duo moves into the role of asset's *digital foundation* (see capsule above). Once again, the asset will sport its own stable of servers, databases, and PC hard drives; however, like the project itself, the naia instances that are tied directly and indirectly to the physical configuration of the asset must be linked back to the *digital foundation*.

---

* Traceability is the one application within a project that could be implemented as a distributed ledger technology solution.

# Managing a Binary Delivery

## *The Importance of the Directrix*

A capital project that has been transformed into a binary delivery framework requires a rethink in how it is managed. The extents of the visibility of the project's naia by the participants create a danger that so many moving parts could end up in a chaotic mess when everyone can dabble into everything. Also, progress tracking is utterly changed from the orthodox approach of periodic updates perpetually out of sync with the actual state of progress at any particular time. Within the binary framework, tracking is instantly available in real time, which enables managers and task leaders to tweak their respective execution schemes on the fly—which means more time spent fine tuning the work. Finally, the availability of powerful *numer* tools to analyze the progress, in real time again, for trends and potential problems, gives the project manager a much greater control over the overall execution scheme, despite the greater number of people looking into how things progress.

For this reason, the old ways of orchestrating the project team by labor types (engineering, procurement, construction, etc.) must be abandoned in favor of the *feature space* emerging from the physical configuration of the asset. By the same token, the oversight structure of the project management team (PMT) needs to be adapted to the *feature space*. The adaption is readily made by implementing the *directrix* principle discussed in Chapter 3. To each installation, system, and component making up the physical configuration of the asset, there corresponds one, and only one, individual assigned a direct accountability mandate. Once the accountability assignments have been nailed down, the probate and responsible mandates are assigned in turn, as shown schematically in Figure 9.5.

The reader will notice the absence of any mention of the digital construct or other binary systems and tools. When the digital construct is limited to the project delivery organization only, its management falls under the purview of the project delivery framework, illustrated in Figure 9.6. If the construct is already part of the owner's organization, its management is folded under the aegis of the Construct Manager (see Figure 7.1). Note that this dual framework-PMT organizational structure borrows from the discussion on team structures and staffing found in Chapters 13 and 14 of *Investment-Centric Project Management*.

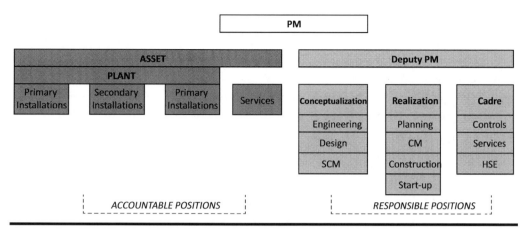

**Figure 9.5   The binary project management team.**

The Framework manages the risks and constraints *external* to the project.

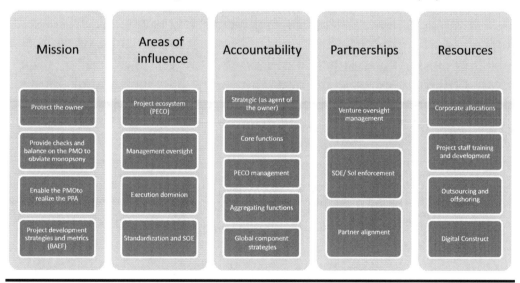

**Figure 9.6   The binary project delivery framework.**

## *The Final Frontier*

Undertaking a project digital management transformation will confront the organization with many hurdles and challenges that are typical of a corporate digital transformation. Making the case for it, planning it, and implementing it must perforce follow the logic of the chapters of this book. At the project level, old hands and newbies alike will struggle with the notion of having the product of their work seen potentially by all. Forcing them to change their content creation habits that are endemic pre-transformation processes (spreadsheets and text documents galore existing on a plethora of PC hard drives) will be challenging—people can get intensely personal with the office tools with which they are most comfortable. But their resistance should wither away quickly once they experience firsthand the efficiency and productivity gains that will come from using digitalized *plik* and *numer* solutions.

One change is necessary to the management philosophy: the concepts of direct accountability and the *directrix* must be implemented from the outset and embedded into the delivery organization's reporting structure. Direct accountability and physical configuration of the asset go hand in hand; maintaining control over the flood of naia instances is only possible through the *directrix*. And the banishment of dabblers, profit killers par excellence, must follow rapidly once the project is underway.*

Supply chain partners (vendors, consultants, fabricators, constructors, inspectors, etc.) may prove less resistant to the transformation when its net effect on them is to streamline datum transaction times and costs. The source of resistance may come from an entirely different realm of privacy and content protection. Companies aren't so willing as people to give up their digital ownership rights for the mere convenience of an app. They will need stalwart assurances that the naia instances that they share electronically with the project team will: 1) be absolutely protected against unauthorized transactions once they are in the team's digital hands; 2) unconditionally retain their digital ownership rights; 3) be tightly controlled in terms of distribution, usage, and copyright protection; 4) be destroyed immediately upon request by the vendor, when such action is contractually permitted; and 5) receive the appropriate training and technical support from the project team on how to interface with the project's digital construct.

The realm of regulators will most likely be the one to resist, if not refuse, getting on board the team's digital transformation. Government entities of this kind are notoriously archaic in their approaches to business process efficiency, while remaining staunchly enamored with paper. Hence, the most likely scenario for the project team is the one that forces the team to forgo

---

* Chapter 5 of *Investment-Centric Project Management* is dedicated to the topic of accountability, which is anathema to dabblers and meddlers.

most of the benefits and efficiencies of the digital transformation and retrograde back to a wasteful transaction schema ordained by the regulatory bodies involved. This is one battle that cannot be won from the outside.

# Bibliography

Keays, Steven J. *Investment-Centric Project Management: Advanced Strategies for Developing and Executing Successful Capital Projects.* J. Ross Publishing, Plantation, FL, 2017, 419 pages.

Keays, Steven J. *Investment-Centric Innovation Project Management: Winning the New Product Development Game.* J. Ross Publishing, Plantation, FL, 2018, 309 pages.

# Chapter 10

# Conclusion

*The journey of a thousand steps... never ends.*

**A digital take on Lao Tzu's famous quote**

## The Tectonic Shifts Are Marching

### Nothing Is Inevitable, until It Is Done

On this day, 2nd March 2019, in the comfortable coffee shop where the bulk of this book was written, an unexpected realization is dawning on your author: the digital transformation that is sweeping across countries and industries and populations has already achieved the status of routine news. The phenomenal advances of technologies that were, just two years ago the stuff of news headlines, no longer occupy the prime real estate of newspapers, magazines, and online media. The stunning news—at least to this author —that Suncor, the Canadian hegemon in oilsands mining, was moving ahead with its plans to replace all its human-driven gargantuan hauling trucks with autonomous ones, barely rippled through the national media. The subject is now finding niche audiences specific to the particular industry, such as *Bits, Bytes, and Barrels* by Geoffrey Cann and Rachael Goydan[*] to name but one. As with all other revolutionary paradigms that cross the chasm from bleeding edge concept to everyday manifestation, the fact that

---

[*] Cann, Geoffrey; Goydan, Rachael. *Bits, Bytes, and Barrels: The Digital Transformation of Oil and Gas.* MadCann Press, 2019, 209 pages.

digital transformations are now part of the routine lingua of business, academic, and government circles cements the fact that the digital paradigm is irreversibly moving forward. We have left the realms of what-if and what-could-be of the last five years to land smack in the middle of "it's happening now, at all scales". The Symbiocene age is upon us, surreptitiously perhaps, but unavoidably. Welcome to your new life...

Business is, as usual, leading the way in spectacular fashion, sometimes publicly, sometimes quietly, but always unflinchingly. Multinational giants weaving digital threads into their supply chains, their business models, and into their future profitability. Surprisingly, some governments have proven themselves ominously eager to get in the game; unnervingly, the leading actors are the dictatures and tyrannies of the world that are going all out, guns ablaze, to deploy vast networks of digital tools in the service of autocracy, coerced speech, and freedom curtailments. Mass surveillance is the killer government app. The Orwellian nightmare was foreseeable long before the enabling technology was. No technological revolution ever came about without its share of nefarious effects and evil intents. Only the Fates know where humanity's faith in technology will land: for the good of all or the benefits of the evil few.

## *You Have Control over Your Destiny*

All is not doom and gloom. Back here in the daily grind, people and organizations are just happy to go about their own business for their own benefit. At this—broadest—scale, the impact of the Symbiocene age has yet to reverberate ubiquitously. But its tremors are starting to be felt by nearly everyone: the day of reckoning is fast approaching. Those who steadfastly refuse to ride the tsunami wave will be swept away like economic flotsam hinted in Figure 3.1. There is no need to fear the new Age, nor to look upon the consequences of digital transformations with apprehension. On the contrary, there is virtually nothing but upsides for those who acquiesce to reality as it is now and plan to make it work to their benefit. It bears repeating the message of this book: digital transformations offer people and organizations fantastic opportunities to improve their lot, enrich their existence, and remain in the driver's seat. There is a sweat irony in this prospect: that the stupendous powers of the digital hegemon can readily be turned in the race for domination between man and machine. These powers are there for taking, by humans, to enfeoff technology to their whim. The rise of the machine at the expense of humanity remains nothing but a utopian fallacy. This book offers the reader a certainty of a forward path on the journey to a digital management transformation. It is, in essence, a comprehensive war plan to conquer the technological terrain and corral its myriad solutions unto a path of human subservience.

> ## LANGUAGE
>
> The case against machines taking over the world when they are able to communicate among themselves without human intermediaries is debunked gleefully by Sloman and Ferbach, in *The Knowledge Illusion.** Whereas machines have long achieved unassailable supremacy in the realm of computing prowess, humans remain solely at the top of the pyramid of actionable reasoning through their sublime ability to communicate intelligently beyond strict rules of grammar, pronunciation, or semantics. Computers will never get the hint, chuckle at a joke, or pick-up on the double and triple entendres of covert speech between conniving humans. No AI could ever hope to defeat a chess champion when the pieces are allowed to randomly change their rules of movement on the board.

Of course, the path of one reader will be unique from other readers intent on their own transformation objectives. Each one will be constrained by their organization's usual culprits: limited budgets, limited resources, lack of time, lack of opportunities without disrupting the flow of business. Readers will rapidly discover the limitations of one's transformation potential, given their starting point, the state of their operating status quo, and the extent of their existing digital footprint. In most instances, the latter will be self-limiting to the point where the only viable transformation will be limited to modernizing one's *plik* systems, processes, and assets. Which, by the way, should be the first objective of *all* digital management transformation missions. Why? Because it yields the most bang for the investment buck in the shortest time (with payback periods measured in weeks or months at most). For instance, the simple implementation of the basic *plik* solutions (cf. Chapter 6, "Low Hanging Fruits") can be carried out by a small team of two to four people over a period of one to two months, *without even needing to go through the development of business calls* (cf. Chapter 5). Gains in productivity and transaction efficiencies will materialize in the weeks following the roll-out of the new *plik* solutions. Virtually all manners of organizations, regardless of size, markets, location, or head count, are candidates for this entry level transformation into the digital world. Even when the scope of the low hanging fruits exceeds the organization's capacity to undertake

---

* Sloman, Steven; Fernbach, Philip. *The Knowledge Illusion: Why We Never Thing Alone.* Riverhead Books, New York, NY, 2017, 304 pages.

them, there are still great valunomic solutions to be pursued on a case-by-case basis. It all depends on what the organization's people transact the most on a daily basis. If your business' naia processes are transacted mainly by spreadsheets, then focus on morphing them into a database schema. If instead the processes are concentrated on Word and PDF documents, then go the template/content generator paradigm first. If it is neither and the bulk of your transactions are via mobile devices, your path is toward the Cloud. And if you deal with specialized software (engineering modeling, financial modeling, seismic visualization, power grid control, sea shipping routing, etc.), then a 3D kernel is your future. It really does not matter what the circumstances of your organization are; whatever they are, take it for granted that profitable opportunities exist today, in the here and now, to change your analog processes and procedures into digitalized mechanics and mechanisms that will cut operating costs, increase revenues, or heighten profits.

Evidently, what is valid for the *plik* goose is also valid for the *numer* gander as well. Innumerable organizations big and small depend on *numer*-type processes and procedures to carry out their daily business, interact with supply chain partners, or liaise with regulators of one form or another. It is not necessary, nor is it advisable in some cases, to pursue an extreme digitalization of these transaction requirements. Going digital does not imply an unconditional embrace of all that is high-end living in the Cloud, existing on an IoT, or ruled by AI. Going digital means, first and foremost, figuring out what the status quo is and educing from it the bottlenecks and inefficiencies that are crippling your efficiencies or eating your profit lunch. The lowest handing fruit is of course the transaction of a naia instance by hand; there is nearly always a valunomic *numer* equivalent mechanism for the job. Targeting one's low hanging *numer* fruits is the safest way into a digital transformation, no matter the organization. It is like dipping your analog toe into the digital lake and finding out that it is the perfect temperature to soak in.

From low hanging fruits, the path forward points toward the advanced *plik* and *numer* solutions. However, the reader is advised to pause before swimming from the digital lake to the numeric ocean of a full-blown digital transformation. The appeal of the low hanging fruit is tampered by concerns for *interoperability*. Adopting digitalizing solutions willy-nilly, without regard to the bigger picture of the organization's strategic objectives, carries the risk of saddling the organization with a plethora of technological orphans incapable of being integrated into a functioning nodal network. The reader is cautioned against favoring the promise of immediate benefits at the expense of future ROI. This is where the execution philosophy conveyed by the book's chapter sequence comes to the fore. The prudent, savvy would-be transformer advocate will be best served by focusing on defining a transformation strategy first. Resist the urge to buy the latest shiny solution making the media rounds. Concentrate instead on the state of your organization's analog status quo (Chapter 4) and quantify what is hurting your bottom line. Then, enunciate in concise terms what you want the organization *to be able to do if you could make it happen with technology.* These are the *model calls* that will quantify the *requirements* and *strategic objectives* of the organization (Chapter 5). Next, you want to quantify the *functional, operating, and performance specifications* of the solutions to be implemented to satisfy the *model calls*—setting the technical foundation of the *digital construct*—discussed at length in Chapter 6. Finally, having thus developed a comprehensive transformation execution strategy, you will undertake the roll-out of the transformation from a position of strength (Chapter 7). This transformation will have effectively changed the physical and algorithmic physiognomy of the organization. This is not the end, however. Next comes what is likely to be most arduous task of the entire journey: the transformation of the human interactions with this new digitalized scheme.

Why is it the most arduous? Because people are notoriously resistant to change, especially when it threatens their perception of their importance to the organization. People are not oblivious to the main themes of the Symbiocene age. Their understanding of the prowess of digital technologies may be superficial or even erroneous but the message that will inevitably remain is the threat to their employment. To the casual consumer of news, the only thing that they hear consistently is the replacement of humans by machines and software. Hence, the dilemma faced by every manager, every business leader, every corporate chieftain: your grandiose vision for a profitable digital future will fall on deaf ears if your employees see right through you, to the only consequence that matters *to them*: losing their jobs. If you cannot assuage this feral instinct *before anything more is said*, you will fail; employees will never go along with a management plan that ultimately tosses them out on the street. At best, they will resist your efforts at every turn. At worst, they will actively, stealthily seek to sabotage them—a far greater risk to the organization's existence than executives are willing to admit.

Corporate hell hath no furry like a workforce scorned.

This fear brings us full circle to the magnificent irony of all digital transformations: ultimately, their success has nothing to do with technological prowess, whizzbang gadgetry, or computational miracles. The ONLY thing that matters, in the end, is the human factor. Which means that those who are called to withstand a digital transformation, willing or not, must be able to see what is in it for them in the long run. There will always be losers (compassionately, not judgementally), unfortunately; but as long as there are a greater number of winners than losers, the transformation stands a good chance of succeeding.

That is why this book calls it a digital *management* transformation, instead of the now-generic expression *digital transformation*. It is about people, not things. Which implies management, not technocrats.

**Calgary, Alberta**
*16 November 2019*

# Bibliography

Sloman, Steven; Fernbach, Philip. *The Knowledge Illusion: Why We Never Thing Alone*. Riverhead Books, New York, NY, 2017, 304 pages.

# Index